Continuous Distributions in Engineering and the Applied Sciences – Part I

Synthesis Lectures on Mathematics and Statistics

Editor
Steven G. Krantz, *Washington University, St. Louis*

Continuous Distributions in Engineering and the Applied Sciences – Part I
Rajan Chattamvelli and Ramalingam Shanmugam
2021

Monte Carlo Methods: A Hands-on Computational Introduction Utilizing Excel
Sujaul Chowdhury
2021

Crowd Dynamics by Kinetic Theory Modeling: Complexity, Modeling, Simulations, and Safety
Bouchra Aylaj, Nicola Bellomo, Livio Gibelli, and Damián Knopoff
2020

Probability and Statistics for STEM: A Course in One Semester
E.N. Barron and J.G. Del Greco
2020

An Introduction to Proofs with Set Theory
Daniel Ashlock and Colin Lee
2020

Discrete Distributions in Engineering and the Applied Sciences
Rajan Chattamvelli and Ramalingam Shanmugam
2020

Affine Arithmetic Based Solution of Uncertain Static and Dynamic Problems
Snehashish Chakraverty and Saudamini Rout
2020

Time Fractional Order Biological Systems with Uncertain Parameters
Snehashish Chakraverty, Rajarama Mohan Jena, and Subrat Kumar Jena
2020

Fast Start Advanced Calculus
Daniel Ashlock
2019

Fast Start Integral Calculus
Daniel Ashlock
2019

Fast Start Differential Calculus
Daniel Ashlock
2019

Introduction to Statistics Using R
Mustapha Akinkunmi
2019

Inverse Obstacle Scattering with Non-Over-Determined Scattering Data
Alexander G. Ramm
2019

Analytical Techniques for Solving Nonlinear Partial Differential Equations
Daniel J. Arrigo
2019

Aspects of Differential Geometry IV
Esteban Calviño-Louzao, Eduardo García-Río, Peter Gilkey, JeongHyeong Park, and Ramón
Vázquez-Lorenzo
2019

Symmetry Problems. The Navier–Stokes Problem.
Alexander G. Ramm
2019
An Introduction to Partial Differential Equations
Daniel J. Arrigo
2017

Numerical Integration of Space Fractional Partial Differential Equations: Vol 2 –
Applicatons from Classical Integer PDEs
Younes Salehi and William E. Schiesser
2017

Numerical Integration of Space Fractional Partial Differential Equations: Vol 1 –
Introduction to Algorithms and Computer Coding in R
Younes Salehi and William E. Schiesser
2017

Aspects of Differential Geometry III
Esteban Calviño-Louzao, Eduardo García-Río, Peter Gilkey, JeongHyeong Park, and Ramón Vázquez-Lorenzo
2017

The Fundamentals of Analysis for Talented Freshmen
Peter M. Luthy, Guido L. Weiss, and Steven S. Xiao
2016

Aspects of Differential Geometry II
Peter Gilkey, JeongHyeong Park, Ramón Vázquez-Lorenzo
2015

Aspects of Differential Geometry I
Peter Gilkey, JeongHyeong Park, Ramón Vázquez-Lorenzo
2015

An Easy Path to Convex Analysis and Applications
Boris S. Mordukhovich and Nguyen Mau Nam
2013

Applications of Affine and Weyl Geometry
Eduardo García-Río, Peter Gilkey, Stana Nikčević, and Ramón Vázquez-Lorenzo
2013

Essentials of Applied Mathematics for Engineers and Scientists, Second Edition
Robert G. Watts
2012

Chaotic Maps: Dynamics, Fractals, and Rapid Fluctuations
Goong Chen and Yu Huang
2011

Matrices in Engineering Problems
Marvin J. Tobias
2011

The Integral: A Crux for Analysis
Steven G. Krantz
2011

Statistics is Easy! Second Edition
Dennis Shasha and Manda Wilson
2010

Lectures on Financial Mathematics: Discrete Asset Pricing
Greg Anderson and Alec N. Kercheval
2010

Fast Start Advanced Calculus
Daniel Ashlock
2019

Fast Start Integral Calculus
Daniel Ashlock
2019

Fast Start Differential Calculus
Daniel Ashlock
2019

Introduction to Statistics Using R
Mustapha Akinkunmi
2019

Inverse Obstacle Scattering with Non-Over-Determined Scattering Data
Alexander G. Ramm
2019

Analytical Techniques for Solving Nonlinear Partial Differential Equations
Daniel J. Arrigo
2019

Aspects of Differential Geometry IV
Esteban Calviño-Louzao, Eduardo García-Río, Peter Gilkey, JeongHyeong Park, and Ramón
Vázquez-Lorenzo
2019

Symmetry Problems. The Navier–Stokes Problem.
Alexander G. Ramm
2019

An Introduction to Partial Differential Equations
Daniel J. Arrigo
2017

Numerical Integration of Space Fractional Partial Differential Equations: Vol 2 –
Applicatons from Classical Integer PDEs
Younes Salehi and William E. Schiesser
2017

Numerical Integration of Space Fractional Partial Differential Equations: Vol 1 –
Introduction to Algorithms and Computer Coding in R
Younes Salehi and William E. Schiesser
2017

Continuous Distributions in Engineering and the Applied Sciences – Part I

Rajan Chattamvelli and Ramalingam Shanmugam

ISBN: 978-3-031-01302-7   paperback
ISBN: 978-3-031-02430-6   ebook
ISBN: 978-3-031-00276-2   hardcover

DOI 10.1007/978-3-031-02430-6

A Publication in the Springer series
SYNTHESIS LECTURES ON MATHEMATICS AND STATISTICS

Lecture #38
Series Editor: Steven G. Krantz, *Washington University, St. Louis*
Series ISSN
Print 1938-1743   Electronic 1938-1751

Continuous Distributions in Engineering and the Applied Sciences – Part I

Rajan Chattamvelli
VIT University, Vellore, Tamil Nadu

Ramalingam Shanmugam
Texas State University, San Marcos, Texas

SYNTHESIS LECTURES ON MATHEMATICS AND STATISTICS #38

ABSTRACT

This is an introductory book on continuous statistical distributions and its applications. It is primarily written for graduate students in engineering, undergraduate students in statistics, econometrics, and researchers in various fields. The purpose is to give a self-contained introduction to most commonly used classical continuous distributions in two parts. Important applications of each distribution in various applied fields are explored at the end of each chapter. A brief overview of the chapters is as follows.

Chapter 1 discusses important concepts on continuous distributions like location-and-scale distributions, truncated, size-biased, and transmuted distributions. A theorem on finding the mean deviation of continuous distributions, and its applications are also discussed. Chapter 2 is on continuous uniform distribution, which is used in generating random numbers from other distributions. Exponential distribution is discussed in Chapter 3, and its applications briefly mentioned. Chapter 4 discusses both Beta-I and Beta-II distributions and their generalizations, as well as applications in geotechnical engineering, PERT, control charts, etc. The arcsine distribution and its variants are discussed in Chapter 5, along with arcsine transforms and Brownian motion. This is followed by gamma distribution and its applications in civil engineering, metallurgy, and reliability. Chapter 7 is on cosine distribution and its applications in signal processing, antenna design, and robotics path planning. Chapter 8 discusses the normal distribution and its variants like lognormal, and skew-normal distributions. The last chapter of Part I is on Cauchy distribution, its variants and applications in thermodynamics, interferometer design, and carbon-nanotube strain sensing. A new volume (Part II) covers inverse Gaussian, Laplace, Pareto, χ^2, T, F, Weibull, Rayleigh, Maxwell, and Gumbel distributions.

Suggestions for improvement are most welcome. Please send them to rajan.chattamvelli@ vit.ac.in.

KEYWORDS

actuarial science, antenna design, arcsine distribution, beta distribution, Brownian motion, civil engineering, communication engineering, electronics, exponential distribution, gamma distribution, Gaussian distribution, process control, reliability, robotics, size-biased distributions, survival analysis, truncated distributions, uniform distribution

Contents

List of Figures

List of Tables

Preface

Continuous distributions are encountered in several engineering fields. They are used either to model a continuous variate (like time, temperature, pressure, amount of rainfall) or to approximate a discrete variate. This book (in two parts) introduces the most common continuous univariate distributions. A comprehensive treatment requires entire volumes by itself, as the literature on these distributions are extensive, and ever-increasing. Hence, only the most important results that are of practical interest to engineers and researchers in various fields are included. Professionals working in some fields usually encounter only a few of the distributions for which probability mass function (PMF), probability density function (PDF), cumulative distribution function (CDF), survival function (SF) (complement of CDF), or hazard rate are needed for special variable values. These appear in respective chapters.

Rajan Chattamvelli and Ramalingam Shanmugam
February 2021

Glossary of Terms

Term	Meaning
ASV	Acceptance Sampling by Variables
CBF	Complete Beta Function
CFR	Constant Failure Rate
ChF	Characteristic Function
CMGF	Central Moment Generating Function
CUNI	Continuous Uniform Distribution
CV	Coefficient of Variation
DAG	Directed Acyclic Graph
DoF	Degrees of Freedom
DSP	Digital Signal Processing
EGD	Exponentiated Gamma Distribution
FWHM	Full Width at Half Maximum
GF	Generating Function
HCD	Half-Cauchy Distribution
HND	Half-Normal Distribution
IBD	Inverted Beta Distribution
IBF	Incomplete Beta Function
IFR	Increasing Failure Rate
IGF	Incomplete Gamma Function
IID	Independently and Identically Distributed
LaS	Location and Scale Distribution
LBM	Linear Brownian Motion
LCC	Line Connecting Centroids
LND	Log-Normal Distribution
LRT	Likelihood-Ratio Tests
MD	Mean Deviation

Term	Meaning
MFP	Mean Free Path
MGF	Moment Generating Function
MLE	Maximum Likelihood Estimate
MoM	Method of Moments Estimation
MVU	Minimum Variance Unbiased
PGF	Probability Generating Function
PVD	Physical Vapor Deposition
RCD	Raised Cosine Distribution
RNG	Random Number Generator
SASD	Standard Arc-Sine Distribution
SBD	Size-Biased Distribution
SCD	Standard Cauchy Distribution
SED	Standard Exponential Distribution
SF	Survival Function
SND	Skew Normal Distribution
SNR	Signal to Noise Ratio
SNSK	Stirling Number of Second Kind
SPD	Standard Probability Distribution
SUD	Standard Uniform Distribution
TBP	True Boiling Point
TTF	Time To Failure
UAV	Unmanned Aerial Vehicles
UND	Unit Normal Distribution (N(0,1))
URN	Unit Random Number
ZTP	Zero-Truncated Poisson Distribution

CHAPTER 1

Continuous Random Variables

1.1 INTRODUCTION

Continuous distributions are encountered in several engineering fields. They are used either to model a continuous variate (like time, temperature, pressure, amount of rainfall) or to approximate a discrete variate. This book (in two parts)[1] introduces the most common continuous univariate distributions. A comprehensive treatment requires entire volumes by itself, as the literature on these distributions are extensive, and ever-increasing. Hence, only the most important results that are of practical interest to engineers and researchers in various fields are included. Professionals working in some fields usually encounter only a few of the distributions for which probability mass function (PMF), probability density function (PDF), cumulative distribution function (CDF), survival function (SF) (complement of CDF), or hazard rate are needed for special variable values. These appear in respective chapters.

1.1.1 CONTINUOUS MODELS

Continuous distributions are more important in industrial experiments and research studies. Measurement of quantities (like height, weight, length, temperature, conductivity, resistance, etc.) on the ratio scale is continuous or quantitative data.

Definition 1.1 The stochastic variable that underlies quantitative data is called a continuous random variable, as they can take a continuum of possible values in a finite or infinite interval with an associated probability.

This can be thought of as the limiting form of a point probability function, as the possible values of the underlying continuous random variable become more and more of fine granularity. Thus, the mark in an exam (say between 0 and 100) is assumed to be a continuous random variable, even if fractional marks are not permitted. In other words, marks can be modeled by a continuous law even though it is not measured at the finest possible granularity level of fractions. If all students scored between 50 and 100 in an exam, the observed range for that exam is of course $50 \leq x \leq 100$. This range may vary from exam to exam, so that the lower limit could differ from 50, and the upper limit of 100 is never achieved (nobody got a perfect 100). This range is in fact immaterial in several statistical procedures.

---

[1]Part I covers rectangular, exponential, beta, arcsine, gamma, cosine, normal, lognormal, and Cauchy distributions; Part II covers inverse Gaussian, Pareto, Laplace, χ^2, Student's T, F, Weibull, Rayleigh, Maxwell, and Gumbel distributions.

All continuous variables need not follow a statistical law. But there are many chance phenomena and physical laws that can be approximated by one of the continuous distributions like the normal law, if not exact. For instance, errors in various measurements are assumed to be normally distributed with zero mean. Similarly, symmetric measurement variations in physical properties like diameter, size of manufactured products, exceedences of dams and reservoirs, and so on, are assumed to follow a continuous uniform law centered around an ideal value θ. This is because they can vary in both directions from an ideal value called its central value.

1.1.2   STANDARD DISTRIBUTIONS

Most of the statistical distributions have one or more parameters. These parameters describe the location (central tendency), spread (dispersion), and other shape characteristics of the distribution. There exist several distributions for which the location information is captured by one, and scale information by another parameter. These are called location-and-scale (LaS) distributions (page 7). There are some distributions called *standard probability distributions* (SPD) for which the parameters are *universally* fixed. This applies not only to LaS distributions, but to others as well.

Definition 1.2   A *standard probability distribution* is a specific member of a parametric family in which all the parameters are fixed so that every member of the family can be obtained by arithmetic transformations of variates.

These are also called "parameter-free" distributions (although location parameter is most often 0, and scale parameter is 1). Examples in univariate case are the standard normal N(0,1) with PDF $f(z) = (1/\sqrt{2\pi})\exp(-z^2/2)$, for which location parameter is 0, and scale parameter is 1; unit rectangular U(0,1), standard exponential distribution (SED) with PDF $f(x) = \exp(-x)$, standard Laplace distribution with PDF $f(x) = \frac{1}{2}\exp(-|x|)$, standard Cauchy distribution with PDF $f(x) = 1/(\pi(1 + x^2))$, standard lognormal distribution with PDF $f(x) = \exp(-(\log(x))^2/2)/(\sqrt{2\pi}\,x)$, and so on. This concept can easily be extended to the bivariate and multivariate probability distributions too. Simple change of origin and scale transformation can be used on the SPD to obtain all other members of its family as $X = \mu + \sigma Z$. Not all statistical distributions have *meaningful* SPD forms, however. Examples are χ^2, F, and T distributions that depend on one or more *degrees of freedom* (DoF) parameters, and gamma distribution with two parameters that has PDF $f(x; a, m) = a^m x^{m-1}\exp(-ax)/\Gamma(m)$. This is because setting special values to the respective parameters results in other distributions.[2] As examples, the T distribution becomes Cauchy distribution for DoF $n = 1$, and χ^2 distribution with $n = 2$ becomes exponential distribution with parameter 1/2.

The notion of SPD is important from many perspectives: (i) tables of the distributions are easily developed for standard forms; (ii) all parametric families of a distribution can be obtained

---

[2]Setting $a = 1$ in gamma distribution results in $f(x; m) = x^{m-1}\exp(-x)/\Gamma(m)$ called one-parameter gamma law, but $m = 1$ results in exponential, and $m = 2$ results in size-biased exponential distributions.

from the SPD form using appropriate variate transformations; (iii) asymptotic convergence of various distributions are better understood using the SPD (for instance, the Student's t distribution tends to the standard normal when the DoF parameter becomes large); and (iv) test statistics and confidence intervals used in statistical inference are easier derived using the respective SPD.

1.1.3   TAIL AREAS

The area from the lower limit to a particular value of x is called the CDF (left-tail area). It is called "probability content" in physics and some engineering fields, although statisticians seem to use "probability content" to mean the volume under bivariate or multivariate distributions. The PDF is usually denoted by lowercase English letters, and the CDF by uppercase letters. Thus, $f(x; \mu)$ denotes the PDF (called Lebesque density in some fields), and $F(x; \mu) = \int_{ll}^{x} f(y)dy = \int_{ll}^{x} dF(y)$, where ll is the lower limit, the CDF (μ denotes unknown parameters). It follows that $(\partial/\partial x)F(x) = f(x)$, and $\Pr[a < X \leq b] = F(b) - F(a) = \int_{a}^{b} f(x)dx$. The differential operator dx, dy, etc. are written in the beginning in some non-mathematics fields (especially physics, astronomy, etc.) as $F(x; \mu) = \int_{ll}^{x} dy f(y)$. Although a notational issue, we will use it at the end of an integral, especially in multiple integrals involving $dxdy$, etc. The quantity $f(x)dx$ is called probability differential in physical sciences. Note that $f(x)$ (density function evaluated at a particular value of x within its domain) need not represent a probability, and in fact could sometimes exceed one in magnitude. For instance, Beta-I(p, q) for $p = 8$, $q = 3$ evaluated at $x = 0.855$ returns 2.528141. However, $f(x)dx$ always represents the probability $\Pr(x - dx/2 \leq X \leq x + dx/2)$, which is in [0,1].

Alternate notations for the PDF are $f(x|\mu)$, $f_x(\mu)$, and $f(x; \mu)dx$, and corresponding CDF are $F(x|\mu)$ and $F_x(\mu)$. These are written simply as $f(x)$ and $F(x)$ when general statements (without regard to the parameters) are made that hold for all continuous distributions. If X is any continuous random variable with CDF $F(x)$, then $U = F(x) \sim U[0, 1]$ (Chapter 2). This fact is used to generate random numbers from continuous distributions when the CDF or SF has closed form. The right-tail area (i.e., SF) is denoted by $S(x)$. As the total area is unity, we get $F(x) + S(x) = 1$. Many other functions are defined in terms of $F(x)$ or $S(x)$. The hazard function used in reliability is defined as

$$h(x) = f(x)/(1 - F(x)) = f(x)/S(x). \tag{1.1}$$

Problem 1.3   Find the unknown C, and the CDF if $f(x) = C(1 - x^2)$, $-1 < x < 1$.

Problem 1.4   Find the unknown C, and the CDF if $f(x) = C(2x - x^3)$, $0 < x < 5/2$.

Problem 1.5   Prove that the hazard function is linear for the PDF $f(x, b) = (1 + bx)\exp(-(x + bx^2/2))$, $x, b > 0$.

Problem 1.6   If X is continuous, prove that $F(x|x > a) = F(x)/(1 - F(a))$, $x > a$.

Problem 1.7 The hazard rate $h(s)$ of pancreatic cancer for an s year old male smoker is $h(s) = 0.027 + 0.00025(s - 40)^2$ for $s \geq 40$. What is the probability that a 40-year-old male smoker survives the age of 70 years without pancreatic cancer?

1.2   MOMENTS AND CUMULANTS

The concept of moments is used in several applied fields like mechanics, particle physics, kinetic gas theory, etc. Moments and cumulants are important characteristics of a statistical distribution and plays an important role in understanding respective distributions. The population moments (also called raw moments or moments around zero) are mathematical expectations of powers of the random variable. They are denoted by Greek letters as $\mu'_r = E(X^r)$, and the corresponding sample moment by m'_r. The zeroth moment being the total probability is obviously one. The first moment is the population mean (i.e., expected value of X, $\mu = E(X)$). There exist an alternative expectation formula for non-negative continuous distributions as $E(X) = \int_{ll}^{ul}(1 - F(x))dx$, where ll is the lower and ul is the upper limit. This takes the simple and more familiar form $E(X) = \int_0^\infty (1 - F(x))dx = \int_0^\infty S(x)dx$ when the range is $(0,\infty)$. Positive moments are obtained when r is a positive integer, negative moments when r is a negative integer, and fractional moments when r is a real number. Alternate expectation formulas exist for higher-order moments as well (see [60]):

$$E(X^r) = \int_0^\infty rx^{r-1}(1 - F(x))dx = \int_0^\infty rx^{r-1}S(x)dx; \quad \text{for} \quad r \geq 1. \tag{1.2}$$

Central moments are moments around the mean, denoted by $\mu_r = E((X - \mu)^r)$. As $E(X) = \mu$, the first central moment is always zero, the second central moment is the variance, the third standardized moment is the skewness, and the fourth standardized moment is the kurtosis.

There are several measures of dispersion available. Examples are mean absolute deviation, variance (and standard deviation (SD)), range, and coefficient of variation (CV) (the ratio of SD and the mean $(s/\bar{x}_n)$ for a sample, and σ/μ for a population). A property of the variance is that the variance of a linear combination is independent of the constant term, if any. Mathematically, $Var(c + bX) = |b|Var(X)$ (which is devoid of the constant c). Similarly, the variance of a linear combination is the sum of the variances if the variables are independent $(Var(X + Y) = Var(X) + Var(Y))$. The SD is the positive square root of variance, and is called *volatility* in finance and econometrics. It is used for data normalization as $z_i = (x_i - \bar{x})/s$ where s is the SD of a sample $(x_1, x_2, \ldots, x_n)$. Data normalized to the same scale or frequency can be combined. This technique is used in several applied fields like spectroscopy, thermodynamics, machine learning, etc. The CV quantifies relative variation within a sample or population. Very low CV values indicate relatively little variation within the groups, and very large values do not provide much useful information. It is used in bioinformatics to filter genes, which are usually combined in a serial manner. It also finds applications in manufacturing engineering, education, and psychology to compare variations among heterogeneous groups as it captures the level of variation relative to the mean.

The inverse moments (also called reciprocal moments) are mathematical expectations of negative powers of the random variable. A necessary condition for the existence of inverse moments is that $f(0) = 0$, which is true for χ^2, F, beta, Weibull, Pareto, Rayleigh, and Maxwell distributions. More specifically, $E(1/X)$ exists for a non-negative random variable X iff $\int_0^\delta (f(x)/x)dx$ converges for some small $\delta > 0$. Although factorial moments can be defined in terms of Stirling numbers, they are not popular for continuous distributions. The absolute moments for random variables that take both negative and positive values are defined as $v_k = E(|X|^k) = \int_{-\infty}^\infty |x|^k f(x)dx = \int_{-\infty}^\infty |x|^k dF(x)$.

The moment generating function (MGF) provides a convenient method to express population moments. It is defined as

$$M_x(t) = E(e^{tx}) = 1 + (t/1!)\mu_1' + (t^2/2!)\mu_2' + \cdots , \tag{1.3}$$

where $\mu_1' = \mu_1 = \mu$ is the mean. A MGF for central moments can be obtained using the relationship $M_{(x-\mu)/\sigma}(t) = \exp(-\mu t/\sigma) M_x(t/\sigma)$. The logarithm of MGF is called cumulant generating function (KGF)

$$K_x(t) = \log(M_x(t)) = (t/1!)k_1 + (t^2/2!)k_2 + \cdots + (t^r/r!)k_r + \cdots , \tag{1.4}$$

so that $k_1 = \mu_1 = \mu$. The characteristic function (also called Fourier–Stieltjes transform) is similarly defined as

$$\phi_x(t) = E(\exp(itx)) = \int_{-\infty}^\infty \exp(itx) f(x)dx. \tag{1.5}$$

They are related as $\phi_x(t) = M_x(it) = \exp(K_x(it))$.

Problem 1.8 Find the mean, variance, and mode of Levy distribution used in economics with PDF $f(x; b, c) = \sqrt{b/(2\pi)}(x - c)^{3/2} \exp(-b/(2(x - c)))$, $x > c$.

Problem 1.9 If the joint PDF of the random locus point on a circle revolving around $(0,0)$ is $f(x, y) = C$ if $x^2 + y^2 \leq r^2$, prove that $C = 1/(\pi r^2)$, and marginal PDF of X is $f(x) = (2/(\pi r^2)) \sqrt{r^2 - y^2}$, $x^2 \leq r^2$, and $f(y) = (2/(\pi r^2)) \sqrt{r^2 - x^2}$, $y^2 \leq r^2$. Find the joint PDF of $T = \sqrt{(x^2 + y^2)}$ and $\theta = \tan^{-1}(y/x)$.

Problem 1.10 If X and Y are identically distributed, but not necessarily independent, is it true that $COV(X + Y, X - Y) = 0$?

1.3 SIZE-BIASED DISTRIBUTIONS

Any statistical distribution with finite mean can be extended by multiplying the PDF or PMF by $C(1 + kx)$, and choosing C such that the total probability becomes one (k is a user-chosen nonzero constant). This reduces to the original distribution for $k = 0$ (in which case $C = 1$). The unknown C is found by summing over the range for discrete, and by integrating for continuous and mixed distributions. This is called size-biased distribution (SBD), which is a special case of weighted distributions. Consider the continuous uniform distribution with PDF $f(x; a, b) = 1/(b - a)$ for $a < x < b$, denoted by CUNI(a, b). The size-biased distribution is $g(y; a, b, k) = [C/(b - a)](1 + ky)$. This means that any discrete, continuous, or mixed distribution for which $\mu = E(X)$ exists can provide a size-biased distribution. As shown in Chapter 2, CUNI(a, b) has mean $\mu = (a + b)/2$. Integrate the above from a to b, and use the above result to get $[C/(b - a)] \int_a^b (1 + ky) dy = 1$, from which $C = 2/[2 + k(a + b)] = 1/(1 + k\mu)$. Similarly, the exponential SBD is $g(y; k, \lambda) = C\lambda(1 + ky) \exp(-\lambda y)$ where $C = \lambda^2/(k + \lambda)$. As another example, the well-known Rayleigh and Maxwell distributions (discussed in Part II) are not actually new distributions, but simply size-biased Gaussian distributions $N(0, a^2)$ with biasing term x, and $N(0, kT/m)$ with biasing term x^2, respectively. Other SBDs are discussed in respective chapters.

We used expectation of a linear function $(1 + kx)$ in the above formulation. This technique can be extended to higher-order polynomials acting as weights (e.g., $(1 + bx + cx^2)$), as also first- or higher-order inverse moments, if the respective moments exist. Thus, if $E(1/(a + bx))$ exists for a distribution with PDF $f(x)$, we could form a new SBD as $g(x; a, b, C) = C f(x)/(a + bx)$ by choosing C so as to make the total probability unity. This concept was introduced by Fisher (1934) [61] to model ascertainment bias in the estimation of frequencies, and extended by Rao (1965, 1984) [113, 115]. More generally, if $w(x)$ is a non-negative weight function with finite expectation $(E(w(x)) < \infty)$, then $w(x) f(x)/E(w(x))$ is the PDF of a weighted distribution (in the continuous case; or PMF in discrete case). It is sometimes called length-biased distribution when $w(x) = x$, with PDF $g(x) = x f(x)/\mu$ (because the weight x acts as a length from some fixed point of reference). As a special case, we could weigh using $E(x^k)$ and $E(x^{-k})$ when the respective moments of order k exists with PDF $x^{\pm k} f(x)/\mu'_{\pm k}$. This results in either distributions belonging to the same family (as in χ^2, gamma, Pareto, Weibull, F, beta, and power laws), different known families (size-biasing exponential distribution by $E(x^k)$ results in gamma law, and uniform distribution in power law), or entirely new distributions (Student's T, Laplace, Inverse Gaussian, etc.). Absolute-moment-based weighted distributions can be defined when ordinary moments do not exist as in the case of Cauchy distribution (see Section 9.4, page 124). Fractional powers can also be used to get new distributions for positive random variables. Other functions like logarithmic (for $x \geq 0$) and exponential can be used to get nonlinearly weighted distributions. The concept of SBD is applicable to classical (discrete and continuous) distributions as well as to truncated, transmuted, exponentiated, skewed, mixed, and other extended distributions.

1.4 LOCATION-AND-SCALE DISTRIBUTIONS

The LaS distributions are those in which the location information (central tendency) is captured in one parameter, and scale (spread and skewness) is captured in another.

Definition 1.11   A parameter θ is called a location parameter if the PDF is of the form $f(x \mp \theta)$, and a scale parameter if the PDF is of the form $(1/\theta) f(x/\theta)$.

Most of the LaS distributions are of the continuous type. Examples are the general normal, Cauchy, and double-exponential (Laplace) distributions. If μ is the mean and σ is the standard deviation of a univariate random variable X, and $Z = (X - \mu)/\sigma$ results in a standard distribution (devoid of parameters; see Section 1.1.2 in page 2), we say that X belongs to the LaS family. This definition can easily be extended to the multivariate case where μ is a vector and Σ is a matrix so that $Z = (X - \mu)'\Sigma^{-1}(X - \mu)$ is in standard form. Sample data values are standardized using the transformation $z_k = (x_k - \overline{x})/s_x$, where s_x is the sample standard deviation, which can be applied to samples from any population including LaS distributions. The resulting values are called z-values or z-scores.

Write the above in univariate case as $X = \mu + \sigma Z$. As σ is positive, a linear transformation with positive slope of any standard distribution results in a location-scale family for the underlying distribution. When $\sigma=1$, we get the one-parameter location family, and when $\mu=0$, we get the scale family. The exponential, gamma, Maxwell, Pareto, Rayleigh, Weibull, and half-normal are scale-family distributions. The CDF of X and Z are related as $F((x - \mu)/\sigma) = G(x)$, and the quantile functions of X and Z are related as $F^{-1}(p) = \mu + \sigma G^{-1}(p)$. For X continuous, the densities are related as $g(x) = (1/\sigma) f((x - \mu)/\sigma)$. Maximum likelihood estimates (MLE) of the parameters of LaS distributions have some desirable properties. They are also easy to fit using available data. Extensions of this include log-location-scale (LLS) distributions, nonlinearly transformed LaS distributions (like trigonometric, transcendental, and other functions of it, etc.) (Jones (2015) [80], Jones and Angela (2015) [81]).

1.5 TRUNCATED DISTRIBUTIONS

There are four types of truncations called left-truncation, right-truncation, symmetric double-truncation, and asymmetric double-truncation. As "symmetric" and "asymmetric" naturally implies that both tails are involved, they are simply called symmetric-truncation and asymmetric-truncation. The concept of truncation is applicable to both symmetric and asymmetric laws. The PDF of a truncated distribution is obtained by dividing the un-truncated (original) PDF by the remaining area (in the continuous case or remaining sum of probabilities in discrete case) enclosed within the new limits. As the total area is unity, this is the same as one minus left out (truncated) area. Symbolically, $g(x; c) = f(x)/(1 - F(c))$, for $x > c$ if c is the truncation point in the left tail, $g(x; c) = f(x)/(1 - S(c))$, for $x < c$ if c is in the right tail, $g(x; c, d) = f(x)/(F(d) - F(c))$ if c is in the left tail, and d is in the right tail ($c < d$). It is convenient to use displacement from the mean for double truncation of symmetric laws as $g(x; c, \mu) = f(x)/(F(\mu + c) - F(\mu - c))$ for $\mu - c < x < \mu + c$.

1.6   TRANSMUTED DISTRIBUTIONS

The CDF of a transmuted distribution is given by

$$G(x; \lambda) = (1 + \lambda) \, F(x) - \lambda \, F^2(x), \tag{1.6}$$

from which the PDF follows by differentiation as

$$g(x; \lambda) = (1 + \lambda) f(x) - 2\lambda f(x) F(x) = f(x)[(1 + \lambda) - 2\lambda F(x)]. \tag{1.7}$$

Consider two or more distributions with the same sample space, and respective CDF $F_k(x)$. For any pair of distributions i and j, form the CDF $G_{ij}(u) = F_j(F_i^{-1}(u))$ and $G_{ji}(u) = F_i(F_j^{-1}(u))$. As $u \in [0,1]$ and the outer CDF maps values to the same range, both of them map unit intervals to itself. This results in transmuted distributions.

1.7   OTHER EXTENSIONS

Classical techniques for extending distributions include convolution, scaling, linear combinations, products, truncations, and nonlinear transformations like inverse $(1/X)$, square, square-root, etc. Introducing one or more new parameters to an existing family of distribution can bring more flexibility for data modelers as a variety of distributional shapes can be incorporated by varying those extra parameters. One example is the Azzalini skew normal family discussed in Chapter 8, that introduces a new parameter to the Gaussian family as $f(x; \lambda) = 2\phi(x)\Phi(\lambda x)$, where $\phi(x)$ is the PDF and $\Phi(x)$ is the CDF of standard normal distribution, and λ is a real number.

The CDF technique can be used to find the PDF of transformed random variables in some of these cases. As examples, if $Y = X^a$, then $G(y) = \Pr[Y \leq y] = \Pr[X^a \leq y] = \Pr[X \leq y^{1/a}]$ $= F(y^{1/a})$. Now differentiate to get the PDF as $g(y) = (1/a)y^{1/a-1} f(y^{1/a})$. This reduces to $G(y) = F(\sqrt{y})$ for $a = 1/2$, and $H(y) = F(y^2)$ for $a = 2$. If the distribution of X is symmetric around zero, $H(y) = F(Y^2 \leq y) = \Pr[-\sqrt{y} \leq Y \leq \sqrt{y}] = F(\sqrt{y}) - F(-\sqrt{y})$. Similarly, the distribution of $Y = 1/X$ can be found using CDF method as $\Pr[Y \leq y] = \Pr[1/X \leq y] = \Pr[X \geq 1/y] = 1 - F(1/y)$.

There exist many other popular extensions of distributions like exponentiated family (EF) of distributions, Marshall-Olkin extended (MOE) family, McDonald generalized (McG) family, Kumaraswamy family, etc. Let $F(x)$ be the CDF of a baseline distribution. Then an EF of distributions can be defined using $G(x) = F(x)^\alpha$ as CDF where $\alpha > 0$ is a positive real number, or as $G(x) = 1 - S(x)^\alpha$ where $S(x)$ is the SF. Differentiate w.r.t. x to get the PDF as $g(x) = \alpha F(x)^{\alpha-1}$. Azzalini's skew symmetric family mentioned above is defined as

$f(x) = 2g(x)G(\lambda x)$ where λ is the skewness parameter, and $g(x), G(x)$ are arbitrary continuous PDF and CDF. As $G(0) = 1/2$ for zero-centered symmetric distributions, this reduces to the baseline distribution for $\lambda = 0$. The Marshall and Olkin (1997) [99] method uses the transformation

$$f(x;c) = cg(x)/[1 - (1 - c)S(x)]^2, \quad S(x) = 1 - F(x), \tag{1.8}$$

for $c \in (0, \infty)$, which reduces to the baseline family for $c = 1$. The beta generated method of Eugene, Lee, and Famoye (2002) [59] is defined in terms of CDF as $G(x) = \int_0^{F(x)} b(t)dt$, where $b(t)$ is the PDF of a beta random variable and $F(x)$ is the CDF of any random variable. This was extended by Alzaatreh, Lee, and Famoye (2013) [6], Aljarrah, Lee, and Famoye (2014) [4] to T-X family of distributions by replacing $b(t)$ by an arbitrary continuous random variable. A McDonald extension uses an exponentiation technique (by raising the variable to a power) to get new distributions. Chattamvelli (2012) [36] introduced the Bernoulli extended family by replacing x in $p^x(1-p)^{1-x} = (1-p)(p/(1-p))^x$ by any continuous CDF or SF, and multiplying by corresponding PDF with an appropriate normalizing constant so that $(p/(1-p))^{F(x)}$ acts as a weight function. Mahdavi and Kundu (2017) [96] used the $\alpha-$ power transformation (APT) method to define a new CDF as $G(x;\alpha) = (\alpha^{F(x)} - 1)/(\alpha - 1)$ for $\alpha \neq 1$. The PDF follows by differentiation as $g(x;\alpha) = (\frac{\log(\alpha)}{\alpha-1} f(x;\alpha) \alpha^{F(x)})$. The Kumaraswamy family is $g(x;a,b) = abF(x)^{b-1}[1 - F(x)^b]^{a-1} f(x)$ where a, b are real constants.

See Alzaatreh, Lee, and Famoye (2013) [6] and Azzalini (1985) [14] for general continuous distributions, Arellano-Valle, Gomez, and Quintana (2004) [10] for skew-normal distributions, Behboodian, Jamalizadeh, and Balakrishnan(2006) [26] for skew-Cauchy distributions, Gupta and Kundu (2009) [70] for weighted exponential distributions, Lemonte et al. (2016) [93] and Rather and Rather (2017) [116] for exponential distributions, McDonald (1995) [97] for beta distribution, Marshall and Olkin (1997) [99] for exponential families, Suksaengrakcharoen and Bodhisuwan (2014) [139] for gamma distributions, and Yalcin and Simsek (2020) [147] for symmetric beta-type distributions.

1.8   MEAN DEVIATION OF CONTINUOUS DISTRIBUTIONS

Finding the mean deviation (MD) of continuous distributions is a laborious task, as it requires lot of meticulous arithmetic work. It is also called the mean absolute deviation or L_1-norm from the mean. It is closely related to the Lorenz curve used in econometrics, Gini index and Pietra ratio used in economics and finance. It is also used as an optimization model for hedging portfolio selection problems [86], for fuzzy multi-sensor object recognition [54], and for minimizing job completion times on computer systems [58]. See Jogesh Babu and Rao (1992) [77] for expansions involving the MD, and Pham-Gia and Hung (2001) [108] for the sampling distribution of MD.

Table 1.1: Summary table of expressions for MD

Distribution Name	Expression for MD	JK Constant	Johnson's Conjecture
Bernoulli	$2pq$	2	$2\mu_2$
Binomial	$2npq\binom{n-1}{\mu-1}p^{\mu-1}q^{n-\mu}, \mu = [np]$	2	$2\mu_2 f_m$
Negative binomial	$2c\binom{k+c-1}{c}q^c p^{k-1}, c = [nq/p]$	2	$2\mu_2 f_m$
Poisson	$2 * \exp(-\lambda)\lambda^{\lfloor\lambda\rfloor+1}/\lfloor\lambda\rfloor!$	2	$2\mu_2 f_m$
Geometric	$(2/q)\lfloor 1/p\rfloor(q^{\lfloor 1/p\rfloor})$	2	$2\mu_2 f_m$
Hypergeometric	$2x(N-k-n+x)\binom{k}{x}\binom{N-k}{n-x}/\left[N\binom{N}{n}\right]$	2	$2\mu_2 f_m[1 + 1/N]$
Uniform	$(b-a)/4$	3	$3\mu_2 f_m$
Exponential	$2/(e\lambda)$	2	$2\mu_2 f_m$
Central χ^2	$e^{-n/2}n^{n/2+1}/[2^{n/2-1}\Gamma(n/2+1)]$	2	$2\mu_2 f_m$
Gamma	$2m^m e^{-m}/(\lambda\Gamma(m))$	2	$2\mu_2 f_m$
Normal	$2\sigma * 1/\sqrt{2\pi} = \sigma\sqrt{2/\pi}$	2	$2\mu_2 f_m$
Arc-sine	$1/\pi$	4	$4\mu_2 f_m$
Cosine	$b(\pi-2)$	2.4424	$2.4424\mu_2 f_m$
Laplace	b	1	$\mu_2 f_m$

Column 3 gives the Johnson–Kamat (JK) constant.

Johnson (1957) [79] surmised that the MD of some continuous distributions can be expressed as $2\mu_2 f_m$ where $\mu_2 = \sigma^2$, and f_m is the probability density expression evaluated at the integer part of the mean $m = \lfloor\mu\rfloor$. This holds good for exponential, normal, gamma, and χ^2 distributions. Kamat [83] extended Johnson's result to several continuous distributions. The multiplier called the Johnson–Kamat (JK) constant is distribution specific (see Table 1.1). The following theorem greatly simplifies the work, and is very helpful to find the MD of a variety of univariate continuous distributions. It can easily be extended to the multivariate case, and for other types of MD like MD from the median and medoid.

Theorem 1.12   Power method to find the Mean Deviation *The MD of any continuous random variable X can be expressed in terms of the CDF as*

$$MD = 2\int_{ll}^{\mu} F(x)dx, \tag{1.9}$$

where ll is the lower limit of X, μ the arithmetic mean, and $F(x)$ the CDF.

Proof. By definition,

$$E|X - \mu| = \int_{ll}^{ul} |x - \mu| f(x) dx \tag{1.10}$$

where *ll* is the lower and *ul* is the upper limit of the distribution. Split the range of integration from *ll* to μ, and μ to *ul*, and note that $|X - \mu| = \mu - X$ for x< μ. This gives

$$E|X - \mu| = \int_{x=ll}^{\mu} (\mu - x) f(x) dx + \int_{x=\mu}^{ul} (x - \mu) f(x) dx. \tag{1.11}$$

Note that $E(X - \mu) = 0$ as $E(X) = \mu$. Expand $E(X - \mu)$ as

$$E(X - \mu) = \int_{ll}^{ul} (x - \mu) f(x) dx = 0. \tag{1.12}$$

As done above, split the range of integration from *ll* to μ and μ to *ul* to get

$$E(X - \mu) = \int_{x=ll}^{\mu} (x - \mu) f(x) dx + \int_{x=\mu}^{ul} (x - \mu) f(x) dx = 0. \tag{1.13}$$

Substitute $\int_{x=\mu}^{ul} (x - \mu) f(x) dx = -\int_{x=ll}^{\mu} (x - \mu) f(x) dx$ in (1.11) to get

$$E|X - \mu| = \int_{x=ll}^{\mu} (\mu - x) f(x) dx - \int_{x=ll}^{\mu} (x - \mu) f(x) dx = 2 \int_{x=ll}^{\mu} (\mu - x) f(x) dx.$$

Split this into two integrals and integrate each of them to get

$$E|X - \mu| = 2 \left[\mu * F(\mu) - \int_{x=ll}^{\mu} x f(x) dx \right], \text{ as F(ll)} = 0. \tag{1.14}$$

Use integration-by-parts to evaluate the second expression:

$$\left\{ x \ F(x) \ |_{ll}^{\mu} - \int_{ll}^{\mu} F(x) dx \right\} = \mu * F(\mu) - ll * F(ll) - \int_{ll}^{\mu} F(x) dx.$$

The $\mu * F(\mu)$ terms cancel out leaving behind

$$E|X - \mu| = 2 \left[x \ F(x) \ |_{ll} + \int_{ll}^{\mu} F(x) dx \right]. \tag{1.15}$$

Here, $x \ F(x) |_{ll} = ll * F(ll)$ means that we are to evaluate the limiting value (from above) of $x * F(x)$ at the lower limit of the distribution. As $F(ll)$ is the cumulative probability (area in univariate case) accumulated *up to* the lower limit, this limit is obviously zero even for U-shaped and reverse J-shaped distributions (like exponential law) with continuities at the extremes. If

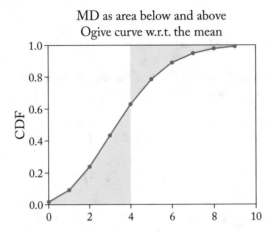

Figure 1.1: MD as sum of CDF and SF.

$\int_{x=ll}^{\mu}(x-\mu)f(x)dx$ in (1.11) is replaced with $\int_{x=\mu}^{ul}(x-\mu)f(x)dx$, an exactly similar steps results in

$$E|X-\mu| = 2\left[\int_{\mu}^{ul} S(x)dx\right], \tag{1.16}$$

where $S(x) = 1 - F(x)$ is the SF. Combining both cases gives another expression for MD as

$$MD = E|X-\mu| = \left[\int_{ll}^{\mu} F(x)dx + \int_{\mu}^{ul} S(x)dx\right]. \tag{1.17}$$

Geometrically, the area below the ogive curve from the lower limit to the mean plus the area above it from the mean to the upper limit is the MD (Figure 1.1). Similarly, twice the area below the ogive curve from the lower limit to the mean, or twice the area above the ogive curve from the mean to the upper limit is also the MD. Thus, we could write the above for all distributions without discontinuities at the extremes as

$$MD = 2\int_{ll}^{\mu} F(x)dx = 2\int_{\mu}^{ul} S(x)dx. \tag{1.18}$$

This representation of MD in terms of CDF is extremely helpful when one needs to evaluate the MD using the CDF or SF (the MD of Beta-I distribution is an integral of the incomplete beta function). This result will be used throughout the book to derive the MD of various distributions. An advantage of this method is that when expressions for tail areas are available in terms of simpler functions (like incomplete beta and gamma functions, orthogonal polynomials, etc.), it is a simple matter to find the MD using the above result. In addition, highly accurate approximations to the CDF of some distributions are available using a power series so that numerical integration techniques can be used to approximate the MD.

1.8.1   NEW MEASURES OF DISPERSION

A new dispersion measure for continuous distributions can be obtained as linear combinations of integrated tail areas as:

$$\Delta = \alpha \int_{ll}^{\mu} F(x)dx + (2-\alpha) \int_{\mu}^{ul} S(x)dx, \tag{1.19}$$

where F(x) is the CDF, S(x) is the SF, ll and ul are, respectively, the lower and upper limits of the distribution, and $0 \leq \alpha \leq 2$ is a constant. This coincides with MD for $\alpha = 0$, $\alpha = 1$, and $\alpha = 2$. Other values of α results in a new measure that asymmetrically combines the tail areas. The right tail is weighted more for $0 < \alpha < 1$, and the left tail for $1 < \alpha < 2$. Replace the mean μ by the median or medoid (nearest integer to the mean) in Eq. (1.19) to get other dispersion measures. Closed form expressions of Δ exist for some distributions. Alternatively, take the difference $\Delta_1 = \alpha \int_{ll}^{\mu} F(x)dx - (1-\alpha) \int_{x=\mu}^{ul} S(x)dx$. It is zero for $\alpha = 0.50$, $-ve$ for $\alpha < 0.5$, and $+ve$ for $\alpha > 0.5$. These results are also extendable to higher dimensional random variables.

1.8.2   A NEW FAMILY OF DISTRIBUTIONS

The result for MD given above can be used to form a new family of distributions. Rewrite Equation (1.18) as

$$(2/MD) \int_{x=ll}^{\mu} F(x)dx = 1. \tag{1.20}$$

Now define a new distribution

$$g_1(x; MD, \mu) = (2/MD)F(x), \quad \text{for} \quad ll \leq x \leq \mu, \tag{1.21}$$

where ll is the lower limit of the distribution. If $G_1(x)$ is the CDF, μ_1 is the mean and MD1 the mean deviation of this new distribution, we could define another new distribution $g_2(x; MD1, \mu_1) = (2/MD1)G_1(x_i)$, and so on. In addition to finding the MD of continuous distributions, this formulation has other important applications in proving convergence of distributions, in central limit theorems. Replace X on the LHS by $S = X_1 + X_2 + \cdots + X_n$. If X_i's are identically distributed continuous random variables, this has mean $n\mu$ so that the relationship becomes

$$E|X_1 + X_2 + \cdots + X_n - n\mu| = \left[\int_{ll}^{\mu} F(x)dx + \int_{\mu}^{ul} S(x)dx \right], \tag{1.22}$$

where F(x) and S(x) are CDF and SF of S. Dividing both sides by n, we see that the first term on the LHS is the arithmetic mean $(X_1 + X_2 + \cdots + X_n)/n$ and RHS has an "n" in the denominator. The RHS tends to zero (because both the integrals either $\rightarrow 1/2$, or their sum $\rightarrow 1$) in the limit as $n \rightarrow \infty$, and the first term on the LHS converges to μ. This provides a simple proof for the asymptotic convergence of independent random variables, which can be extended to other cases. For example, if $g(X_i)$ is a continuous function of X_i with finite mean $v = g(\mu)$, replacing X_i by $g(X_i)$ in (1.22) provides a simple proof on the convergence of $g(X_i)$ to its mean asymptotically.

Similar expressions are available for the mean deviation around the median as

$$E|X\text{-Median}| = \int_0^{\frac{1}{2}} (F^{-1}(1-x) - F^{-1}(x))dx = \int_0^{\frac{1}{2}} (S^{-1}(x) - S^{-1}(1-x))dx. \quad (1.23)$$

As mean = median for a symmetric distribution, $E|X\text{-Median}| = E|X\text{-Mean}|$ for it.

Theorem 1.13   *Variance of continuous distributions as tail areas*
Prove that the variance of a continuous distribution is an integral of tail area.

Proof. We found above that $MD = 2\int_{x=ll}^{\mu} F(x)dx = 2\int_{x=\mu}^{ul} S(x)dx$ where $F(x)$ is the CDF and $S(x)$ is the SF. Equating Johnson's result that $MD = c\mu_2 f_m$ where $\mu_2 = \sigma^2$ and f_m is the probability mass evaluated at the integer part of the mean $m = \lfloor \mu \rfloor$ we get $\mu_2 * c f_m = 2\int_{x=ll}^{\mu} F(x)dx$. Divide both sides by $c f_m$ to get

$$\mu_2 = \sigma^2 = (2/(c*f_m)) \int_{x=ll}^{\mu} F(x)dx = (2/(c*f_m)) \int_{x=\mu}^{ul} S(x)dx. \quad (1.24)$$

An alternate expression given by Jones and Balakrishnan (2003) [82] is

$$\sigma^2 = 2 \int_y \int_{-ll<x<y}^{ul} F(x)[1 - F(y)]dxdy. \quad (1.25)$$

The constant multiplier ($c = 2$) proposed by Johnson may be different for some continuous distributions (e.g., for Laplace distribution $c = 1$). The above result holds even then because the RHS simply gets scaled by c.

1.9   RANDOM NUMBERS

As the name implies, random numbers (RN) are arbitrary numbers generated for a specific purpose. As a simple example, a random digit is one among the 10 digits (0–9). They find applications in computer games, simulation programs, data encryption, operating systems, etc. RNs can be generated from a discrete or continuous population. Continuous RNs from a univariate distribution are generated using an interval $[a, b]$. Here a and b are real numbers where $a < b$, and boundaries are included, meaning that generated numbers could coincide with a and b.

Definition 1.14   Numbers in a specific range that are generated in arbitrary order are called random numbers.

Most of the popular programming languages have built-in random number generators (RNG). Some of the RNGs generate a uniform random number in a specific range (say 0 to

32767) while others generate it in the closed unit interval [0,1]. These are called unit random number (URN) that finds applications in generating RN from arbitrary statistical distributions. If [a,b] is a sub-interval of (0,1), the probability that the next generated random number will lie in it is (b-a). A URN can be scaled to an appropriate interval. For instance, if u is a URN, $x = a + (b - a)u$ is an RN in the range $[a, b]$, and $x = k[2u - 1]$ is an RN in the range $[-k, +k]$ [35]. Put $a = 0$ to get $x = bu$ as an RN in the range $[0, b]$. These are equally likely to lie anywhere in the range. Nonlinear transformation of URN are used to get RNs from some distributions. One example is the one-parameter exponential distribution $\lambda \exp(-\lambda x)$, for which the transformation is $-\log(u)/\lambda$. These are discussed in upcoming chapters.

1.10   DATA OUTLIERS

Outliers in a sample are "extreme data values" that fall significantly away from the rest of the data values. Thus, univariate outliers are unusually off (in either the left (lower) or right (higher) side) from the pattern of variability evident among other data values.

Definition 1.15   A data outlier is a sample value that deviates unusually away from the rest of the data values.

There exist several outlier detection techniques, the simplest of which is using a graphical display of the data. Univariate outliers can also be detected using $\bar{x} \mp ks$ rule, where $\bar{x}$ is the sample mean, s the standard deviation, and k is a constant ≥ 3. This implies that data points that fall beyond $(\bar{x} - 3s, \bar{x} + 3s)$ range can be considered as data outliers for two-tailed distributions. The value of k should be chosen based on the population distribution. For example, the Pareto, Laplace, and Student's t distributions tail off slower than the Gaussian law. Although $k = 3$ can be used to detect outliers from Gaussian distributions, a higher value of k is needed for other *thick-tailed* distributions. If the population parameters are known, we could define an α-outlier as that data point which falls in $1 - \alpha$ critical region. This can be found for each distribution using the tail areas described in Section 1.1.3. Mahalanobis' distance (discussed in Part-II book) is used to detect multivariate outliers when correlation is present among the variables. Outlier data values can influence statistics (functions of sample values) computed from it. Detecting and removing outliers is called data cleansing.

1.11   SUMMARY

Essential concepts on continuous distributions are introduced in this chapter. Many new distributions can be obtained from existing ones using special techniques like size-biasing, truncation, exponentiation, convolution, etc. Several other types of extensions exist as in transmuted, beta-generalized, Kumarasamy, Marshal–Olkin, McDonalds family of distributions. A simple method to find the MD of continuous distributions is discussed, which can be generalized to higher dimensions. A new measure of dispersion that *asymmetrically* combines the tail areas is introduced.

CHAPTER 2

Rectangular Distribution

2.1 INTRODUCTION

As the name implies, this distribution assigns a constant probability to each point in a continuous interval. Thus, the range is always finite (and quite often small in practical applications). The continuous uniform distribution is denoted by CUNI(a, b) or U(a, b) (and discrete uniform distribution by DUNI(a, b)), where the first argument a is the lower, and the second argument $b(> a)$ is the upper limit. It is also called continuous uniform, or simply, uniform distribution because the PDF is shaped like a rectangle with base $(b - a)$ and height $1/(b - a)$, so that the area is $(b - a) \times 1/(b - a) = 1$. The corresponding random variable is denoted by U when the range is unity, and by X otherwise. The PDF of CUNI(a, b) is given by

$$f(x; a, b) = \begin{cases} 1/(b - a) & \text{for} \quad a \leq x \leq b; \\ 0 & \text{otherwise.} \end{cases} \tag{2.1}$$

It is called the standard uniform (or unit uniform) distribution (SUD) when $a = 0$ and $b = 1$, and is written as $U(0, 1)$ or simply U. This has PDF $f(u) = 1$, $0 < u < 1$. The CDF of CUNI(a, b) is obtained by integrating (2.1) from a to x as $F_X(x; a, b) = \int_a^x [1/(b - a)]dy = [1/(b - a)]y|_a^x = (x - a)/(b - a)$. Thus,

$$F(x; a, b) = \begin{cases} 0 & \text{for} \quad x < a; \\ (x - a)/(b - a) & \text{for} \quad a \leq x \leq b; \\ 1 & \text{for} \quad x > b. \end{cases} \tag{2.2}$$

Considered as an algebraic equation, $y = (x - a)/(b - a)$ represents a straight line with slope $1/(b - a)$, and intercept $a/(a - b)$. This line segment is defined only within the interval (a, b). The slope is small when the range $(b - a)$ is large. The slope is large (line is steep) in the limiting case $b \to a$.

This distribution has extensive applications in many applied science fields. Examples are symmetric measurement errors in engineering, errors in physical specifications of manufactured products. These are often due to wear and tear of machinery and parts thereof, non-uniformity of ingredients, variations in environments like temperature and humidity in the manufacturing plant etc. If an *ideal* value is known, and lower and upper tolerance limits are specified, the two-sided symmetric specifications can be represented as [ideal−tolerance, ideal+tolerance]. Thickness distribution of lubricant layers on some products like hard disks, moving belts, etc., as well as magnetic disk substrates, are also approximately uniform distributed. Similarly, round-off errors in numerical calculations that truncate real numbers to nearest k decimal places are assumed to be uniformly distributed.

2.1.1   ALTERNATE REPRESENTATIONS

The range of the distribution is symmetric around the origin (say $-a$ to $+a$), or centered around a fixed constant (say $\theta - 1/2, \theta + 1/2$) in several practical applications:

$$f(x; a, \theta) = \begin{cases} 1/\theta & \text{for} \quad a \leq x \leq a + \theta; \\ 1/(2\theta) & \text{for} \quad a - \theta \leq x \leq a + \theta; \\ 0 & \text{otherwise.} \end{cases} \tag{2.3}$$

For $a = 1$ we get $f(x; a, b) = 1/2$ for $-1 < x < 1$; and for $a = 1/2$ we get $f(u; a, b) = 1$ for $-1/2 < u < 1/2$.

The mean and variance are $\mu = (a + b)/2$, and $\sigma^2 = (b - a)^2/12$, as shown on page 21. Write $\mu = (a + b)/2$ and $\sigma = (b - a)/(2\sqrt{3})$. Cross multiply to get $(a + b) = 2\mu$, and $(b - a) = (2\sqrt{3})\sigma$. Add them to get $b = \mu + \sqrt{3}\sigma$. Subtracting gives $a = \mu - \sqrt{3}\sigma$, from which $(b - a) = (2\sqrt{3})\sigma$. Thus, the PDF becomes

$$f(x; \mu, \sigma) = 1/(2\sqrt{3}\sigma), \quad \mu - \sqrt{3}\sigma \leq x \leq \mu + \sqrt{3}\sigma. \tag{2.4}$$

The CDF is

$$F(x; a, b) = \begin{cases} 0 & \text{for} \quad x < \mu - \sqrt{3}\sigma; \\ 0.50 \left(1 + (x - \mu)/(\sqrt{3}\sigma)\right) & \text{for} \quad \mu - \sqrt{3}\sigma \leq x \leq \mu + \sqrt{3}\sigma; \\ 1 & \text{for} \quad x > \mu + \sqrt{3}\sigma, \end{cases} \tag{2.5}$$

and inverse CDF is $F^{-1}(p) = \mu + \sqrt{3}\sigma(2p - 1)$ for $0 < p < 1$. Put $\sigma = 1$ to get the standardized CUNI$(-\sqrt{3}, +\sqrt{3})$ that has mean zero and variance unity.

Some applications in engineering, theoretical computer science, and number theory use the uniform distribution *modulo k*. This allows the distribution to be extended to the entire real line (because the "mod k" maps all such real numbers to $(0, k)$ range), and are more applicable to discrete uniform distribution. The uniform distribution on a circle has PDF $f(x) = 1/(2\pi)$, for $0 < x \leq 2\pi$.

2.2   RELATED DISTRIBUTIONS

Due to its relationship with many other distributions, it is extensively used in computer generation of random variables. $U(0, 1)$ is a special case of Beta-I(a, b) when $a = b = 1$. If $X \sim U(0, 1)$ then $Y = -\log(X) \sim$ SED (i.e., EXP(1)), and $Y = -2\log(X)$ has a χ_2^2 distribution. If $x_1, x_2, \ldots, x_k$ are independent samples from possibly k different U(0,1) populations,

$P_k = \sum_{j=1}^{k} -2\ln(x_j)$ being the sum of k IID χ_2^2 variates has χ_{2k}^2 distribution. This is called Pearson's statistic in tests of significance [114]. A simple change of variable transformation $Y = (X-a)/(b-a)$ in the general PDF results in the SUD (i.e., U(0,1)). U(0,1) is also related to arcsine distribution as $Y = -\cos(\pi U/2)$ (Chapter 5). If X is any continuous random variable with CDF $F(x)$, then $U = F(x) \sim U[0,1]$.

Example 2.1   Distribution of F(x)
If X is a continuous variate, find the distribution of $U = F(x)$.

Solution 2.2   Consider

$$F(u) = \Pr(U \le u) = \Pr(F(x) \le u) = \Pr(x \le F^{-1}(u)) = F[F^{-1}(u)] = u. \tag{2.6}$$

The CDF of a rectangular distribution CUNI(a,b) is $(x-a)/(b-a)$. Put $a = 0, b = 1$ to get $F(x) = x$. Equation (2.6) then shows that U is an SUD.

   This property can be used to generate random numbers from a distribution if the expression for its CDF (or SF) involves simple or invertible arithmetic or transcendental functions. For example, the CDF of an exponential distribution is $F(x) = 1 - \exp(-\lambda x)$. Equating to a random number u in the range [0,1] and solving for x, we get $1 - e^{-\lambda x} = u$ or $x = -\log(1-u)/\lambda$, using which random numbers from exponential distributions can be generated.

Problem 2.3   If a doctor and a nurse arrive a hospital independently and uniformly during 8 AM and 9 AM, find the probability that the first patient to arrive has to wait longer than 10 min, if consultation is possible only when both the doctor and nurse are in office.

Problem 2.4   If three random variables X, Y, Z are independently and uniformly distributed over (0,1), show that $\Pr(X > YZ) = 3/4$.

Example 2.5   Logarithmic transformation of CUNI Distribution
If X is CUNI(a,b) distributed, find the distribution of $Y = -\log(X-a)/(b-a)$.

Solution 2.6   The CDF of Y is $F(y) = \Pr[Y \le y] = \Pr[(X-a)/(b-a) \ge \exp(-y)] = \Pr[X \ge a + (b-a)\exp(-y)]$. As the CDF of CUNI$(a,b)$ is $(x-a)/(b-a)$, this becomes $1 - \Pr[X \le a + (b-a)\exp(-y)] = 1 - \exp(-y)$. From this the PDF is obtained by differentiation as $f(y) = \exp(-y)$. Hence, $Y \sim$ EXP(1).

Problem 2.7   What is the probability that the length of a randomly selected chord is greater than the side of the equilateral triangle inscribed in the circle?

Problem 2.8   If $X \sim U(0,1)$, find the distribution of $Y = -\log(1-X)$. Hence, prove that $\Pr[Y|X > x_0] = 1 - \log(1-x_0)$ characterizes the $U(0,1)$ distribution.

Problem 2.9   If X and Y are IID $U[0, 1]$, find the distribution of $-2\log(XY)$.

Example 2.10   Reciprocal of a unit rectangular variate   If $X \sim U(0, 1)$, find the distribution of $Y = 1/X$.

Solution 2.11   Let $G(y)$ be the CDF of Y. Then

$$G(y) = \Pr[Y \leq y] = \Pr[1/X \leq y] = \Pr[X \geq 1/y] = 1 - 1/y. \qquad (2.7)$$

Differentiate w.r.t. y to get the PDF as $g(y) = 1/y^2$, for $y \geq 1$.

Problem 2.12   If three points are chosen at random on the circumference of a unit circle, what is the probability that (i) they determine an acute-angled triangle and (ii) they lie on a semi-circle.

Problem 2.13   Find the mean distance of two points randomly chosen on the circumference of a circle.

Example 2.14   Distribution of $Y = \tan(X)$ If $X \sim U(0, 1)$, find the distribution of $Y = \tan(X)$.

Solution 2.15   The PDF of X is $f(x) = 1$, $0 < x < 1$. The inverse transformation is $X = \tan^{-1}(Y)$. This gives $|\partial x/\partial y| = 1/(1 + y^2)$. The range of Y is modified as $\tan(0) = 0$ to $\tan(1) = \pi/4$. Hence the distribution of Y is $f(y) = 1/(1 + y^2)$ for $0 < y < \pi/4$.

Example 2.16   Distribution of $U = \sin(X)$ If X has a CUNI$[-\pi/2, \pi/2]$ distribution, find the distribution of $U = \sin(X)$.

Solution 2.17   The inverse transformation is $x = \sin^{-1}(u)$ so that $|\partial x/\partial u| = 1/\sqrt{1 - u^2}$. When $x = -\pi/2$, $u = \sin(-\pi/2) = -\sin(\pi/2) = -1$. When $x = \pi/2$, $u = \sin(\pi/2) = 1$. As the PDF of X is $1/\pi$, the PDF of U is $f(u) = (1/\pi)/\sqrt{1 - u^2}$, for $-1 < u < +1$.

2.3   PROPERTIES OF RECTANGULAR DISTRIBUTION

This distribution has a special type of symmetry called *flat-symmetry*. Hence, all odd central moments μ_{2r+1} except the first one are zeros. The median always coincides with the mean, and the mode can be any value within the range. As the probability is constant throughout the interval, the range is always finite (and quite often small). As $F(x) = (x - a)/(b - a)$, its inverse is

$$F^{-1}(p) = a + p(b - a), \quad \text{for} \quad 0 < p < 1. \qquad (2.8)$$

A uniform distribution defined in an interval $(c, c + \theta)$ has PDF

$$f(x; \theta) = 1/\theta \quad \text{for} \quad c \leq x \leq c + \theta. \qquad (2.9)$$

Take $c = 0$ to get the standard form $f(x; \theta) = 1/\theta$, $0 < x < \theta$. This is the analogue of the DUNI(N) with probability function $f(x; N) = 1/N$, for $x = 0, 1, 2, \ldots, N - 1$ discussed in Chapter 3 of Chattamvelli and Shanmugam (2020) [42]. The transformation $Y = (b - a) - X$ results in the same distribution. In particular, if $X \sim U(0, 1)$ then $Y = 1 - X \sim U(0, 1)$. This property of $U(0, 1)$ is used in generating random samples from other distributions like the exponential distribution (page 35). Only the extremes of a sample $x_{(1)}$ and $x_{(n)}$ are sufficient to fit this distribution.

Problem 2.18 If $X \sim U[0,1]$, find the distribution of $Y = \exp(X)$, and its variance.

Problem 2.19 If $X \sim U(0, 1)$, find the distribution of $Y = X/(1 - X)$.

2.3.1   MOMENTS AND GENERATING FUNCTIONS

The moments are easy to find using the MGF. The mean is directly obtained as $\mu = [1/(b - a)] \int_a^b x\, dx = [1/(b - a)] \frac{x^2}{2}\big|_a^b = (b^2 - a^2)/[2(b - a)] = (a + b)/2$. Higher-order moments are found using the MGF. Thus,

$$E(X^n) = (1/(n + 1)) \sum_{k=0}^{n} a^k b^{n-k}. \tag{2.10}$$

The MGF is

$$M_x(t) = E(e^{tx}) = \int_{x=a}^{b} [1/(b - a)]e^{tx}\, dx = [1/(b - a)]e^{tx}/t\big|_a^b = (e^{bt} - e^{at})/[(b - a)t]. \tag{2.11}$$

The characteristic function (ChF) is

$$\phi_x(t) = (\exp(ibt) - \exp(iat))/[(b - a)it] \quad \text{for} \quad t \neq 0. \tag{2.12}$$

This reduces to $\sinh(at)/at$ for CUNI$(-a, +a)$.

Moments can be found from the MGF as follows. Consider $e^{bt}/t = 1/t + b + b^2 t/2! + \cdots + b^k t^{k-1}/k! + \cdots$. As $(1/t)$ is common in both e^{bt}/t and e^{at}/t, it cancels out. The second term is $(b - a)/(b - a) = 1$. Thus,

$$(e^{bt} - e^{at})/[(b - a)t] = 1 + \frac{1}{b - a}\left(\sum_{k=2}^{\infty} [(b^k - a^k)/(b - a)]t^{k-1}/k!\right). \tag{2.13}$$

If we differentiate (2.13) $(k - 1)$ times w.r.t. t, all terms below the $(k - 1)^{th}$ term will vanish (as they are derivatives of constants independent of t's) and all terms beyond the kth term will contain powers of t. Only the $(k - 1)^{th}$ term is a constant with a $(k - 1)!$ in the numerator, which cancels out with the $k!$ giving a k in the denominator. By taking the limit as $t \to 0$, we get

$$\mu'_{k-1} = (\partial^{k-1}/\partial t^{k-1})M_x(t)|_{t=0} = (b^k - a^k)/[(b - a)k]. \tag{2.14}$$

Table 2.1: Properties of continuous uniform distribution U(a,b).

Property	Expression	Comments
Range of X	$a \leq x \leq b$	Continuous; finite
Mean	$\mu = (a+b)/2$	Median $= (a+b)/2$
Variance	$\sigma^2 = (b-a)^2/12$	$\sigma^2 = (\mu^2 - ab)/3$
Skewness	$\gamma_1 = 0$	Special symmetry
Kurtosis	$\beta_2 = 9/5$	
MD	$E\lvert X - \mu \rvert = (b-a)/4$	$(\sqrt{3}/2)\sigma = 0.866\sigma$
CV	$(b-a)/[\sqrt{3}(a+b)]$	0.57735 if $a = 0$, $b = 1$
CDF	$(x-a)/(b-a)$	Line sloping up
SF	$(b-x)/(b-a)$	
Moments (μ'_r)	$(b^{r+1} - a^{r+1})/[(b-a)(r+1)]$	
Moments	$\mu_r = [(b-a)/2]^r/(r+1)$	r even
MGF	$(e^{bt} - e^{at})/[(b-a)t]$	
ChF	$(e^{ibt} - e^{iat})/[(b-a)it]$	
PGF	$(t^b - t^a)/[(b-a)\log(t)]$	

SUD results when $a = 0$, $b = 1$.

Putting $k = 2$ gives $\mu_1 = (b+a)/2$. The second moment is obtained by putting $k = 3$ as $\mu'_2 = (b^3 - a^3)/[3(b-a)] = (b^2 - ab + a^2)/3$. From this we get the second central moment as $\mu_2 = (b^2 - ab + a^2)/3 - (a+b)^2/4$. Taking 12 as the LCM of 3 and 4, this simplifies to $\mu_2 = \sigma^2 = (b-a)^2/12$, so that the standard deviation (SD) is $(b-a)/2\sqrt{3}$. The SD reduces to $a/\sqrt{3}$ for CUNI$(-a, +a)$. See Table 2.1 for further properties.

Problem 2.20   If $X \sim U(0,1)$, prove that $Y = -\ln(X)$ is standard exponential. What is the distribution when the logarithm is not to the base e?

Problem 2.21   If $X \sim U(0,1)$, find the distribution of (i) $Y = 1 - \exp(-x)$ (ii) $c * \tan(1/X)$.

Problem 2.22   Suppose a random variable X follows a uniform PDF with support $(-1, +1)$. What is the PDF of $\lvert X \rvert$? Obtain the probability that (i) $\lvert X \rvert > 1/3$ and (ii) $\lvert X \rvert < 1/2$.

Problem 2.23   If $X_1, X_2, \ldots, X_n$ is a random sample of size n from CUNI(a, b), show that $\Pr[\sum_{k=1}^{n} x_k \leq c] = (1/n!)\sum_{j=0}^{c}(-1)^j \binom{n}{j}(c-j)^j$, $0 \leq c \leq n$.

Example 2.24   Distribution of $Y = \log(U/(1-U))$ If $X \sim U(0, 1)$ find the distribution of $Y = \log(U/(1-U))$.

Solution 2.25 Let $F(y)$ be the CDF of Y. Then $F(y) = \Pr[Y \le y] = \Pr[\log(U/(1-U)) \le y] = \Pr[(U/(1-U)) \le \exp(y)]$. Cross multiply and simplify to get $\Pr[U \le \exp(y)/(1+\exp(y))]$. As U is uniform, this becomes $F(y) = \exp(y)/(1+\exp(y))$. Differentiate w.r.t. y to get the PDF as $f(y) = [(1+\exp(y))\exp(y) - \exp(y)\exp(y)]/(1+\exp(y))^2$, which simplifies to $1/(1+\exp(y))^2$.

Example 2.26 MD of rectangular distribution Find the MD of rectangular distribution.

Solution 2.27 By definition, $E|X - \mu| = \int_a^b |x - \mu|/(b-a)dx$. Split the range of integration from "a" to μ and μ to "b", and note that $|X - \mu| = \mu - X$ for $x < \mu$. This gives

$$E|X - \mu| = \int_{x=a}^{\mu} (\mu - x)/(b-a)dx + \int_{x=\mu}^{b} (x - \mu)/(b-a)dx. \tag{2.15}$$

Consider $\int_a^\mu \mu dx - \int_\mu^b \mu dx$. As $\mu = (a+b)/2$, this integral vanishes. What remains is

$$\frac{1}{b-a}\left(\int_\mu^b x\,dx - \int_a^\mu x\,dx\right) = \frac{1}{2(b-a)}(b^2 - \mu^2 - (\mu^2 - a^2)) = \frac{(a^2 + b^2 - 2\mu^2)}{2(b-a)}. \tag{2.16}$$

Substitute the value $\mu = (a+b)/2$ and take 2 as a common denominator to get $\frac{1}{2(b-a)}(b-a)^2/2 = (b-a)/4$. Thus, the mean deviation $E|X - \mu| = (b-a)/4$. This becomes $\theta/2$ for the alternate form of the PDF (in Eq. (2.3)).

Next, we apply Theorem 1.12 (page 10) to verify our result. As the CDF is $(x-a)/(b-a)$, the MD is given by

$$MD = 2\int_{ll}^{\mu} F(x)dx = 2/(b-a)\int_a^c (x-a)dx \text{ where } c = \mu = (a+b)/2. \tag{2.17}$$

The integral $\int_a^c (x-a)dx$ is $(x-a)^2/2|_a^c$. The integral evaluated at the lower limit is zero. As $c = (a+b)/2$, the upper limit evaluates to $(b-a)^2/8$. Substitute in (2.17). One $(b-a)$ cancels out and we get the MD as $(b-a)/4$. This tallies with the above result.

Problem 2.28 If X_1, X_2 are IID CUNI(0,1), find the distribution of $Y_1 = \sqrt{-2\log_e(x_1)}\cos(2\pi x_2)$ and $Y_2 = \sqrt{-2\log_e(x_1)}\sin(2\pi x_2)$ (the inverse transformation being $X_1 = \exp\left[-\frac{1}{2}(y_1^2 + y_2^2)\right]$ and $x_2 = \frac{1}{2\pi}\arctan(y_2/y_1)$).

Problem 2.29 If $X \sim CUNI(a,b)$, find the PDF of $Y = -\log(X)$. If $a = -b$, find the distribution of $Y = X^2$.

Problem 2.30   If $X \sim \text{CUNI}(-(a+b)/2, (a+b)/2)$, find the distribution of $Y = |X|$.

Problem 2.31   If $U \sim \text{CUNI}(0,1)$ find the distribution of (i) $|U - \frac{1}{2}|$, (ii) $1/(1+U)$.

Problem 2.32   If X_1, X_2 are IID with PDF $f(x) = 1/x^2$, $\quad 1 < x < \infty$, find the distribution of $X_1 X_2$.

Problem 2.33   An electronic circuit consists of two independent identical resistors connected in parallel. Let X and Y be the lifetimes of them, distributed as $\text{CUNI}(0, b)$ with PDF $f(x) = 1/b$, $0 < x < b$. Find the distribution of (i) $Z = X + Y$ and (ii) $U = XY$.

2.3.2   TRUNCATED UNIFORM DISTRIBUTIONS

Truncation in the left-tail or right-tail results in the same distribution with a reduced range. Suppose $X \sim \text{CUNI}(a, b)$. If truncation occurs in the left-tail at $x = c$ where $a < c < b$, the PDF is given by

$$g(x; a, b, c) = f(x; a, b)/(1 - F(c)) = (1/(b-a))[1/(1 - (c-a)/(b-a))] = 1/(b-c). \tag{2.18}$$

If truncation occurs at c in the left-tail and d in the right-tail, the PDF is given by

$$g(x; a, b, c, d) = f(x; a, b)/(F(d) - F(c)) \quad = 1/(d - c). \tag{2.19}$$

This shows that truncation results in rectangular distributions.

Example 2.34   Even moments of rectangular distribution   Prove that the kth central moment is zero for k odd, and is given by $\mu_k = (b-a)^k/[2^k(k+1)]$ for k even.

Solution 2.35   By definition, $\mu_k = \frac{1}{b-a} \int_a^b (x - \frac{a+b}{2})^k dx$. Make the change of variable $y = x - (a+b)/2$. For $x = a$, we get $y = a - (a+b)/2 = (a-b)/2 = -(b-a)/2$. Similarly for $x = b$, we get $y = b - (a+b)/2 = (b-a)/2$. As the Jacobian is $\partial y/\partial x = 1$, the integral becomes $\mu_k = \frac{1}{b-a} \int_{-(b-a)/2}^{(b-a)/2} y^k dy$. When k is odd, this is an integral of an odd function in symmetric range, which is identically zero. For k even, we have $\mu_k = \frac{2}{b-a} \int_0^{(b-a)/2} y^k dy = \frac{2}{b-a}[y^{k+1}/(k+1)]|_0^{(b-a)/2} = (b-a)^k/[2^k(k+1)]$, as the constant $2/(b-a)$ cancels out.

Example 2.36   Ratio of independent uniform distributions   If X and Y are IID $\text{CUNI}(0, b)$ variates, find the distribution of $U = X/Y$.

Solution 2.37   Let $U = X/Y, V = Y$ so that the inverse mapping is $Y = V, X = UV$. The Jacobian is $|J| = v$. The joint PDF is $f(x, y) = 1/b^2$. Hence, $f(u, v) = v/b^2$. The PDF of u is obtained by integrating out v. The region of interest is a rectangle of sides $1 \times b$ at the left, and a curve $uv = b$ to its right. Integrating out v, we obtain $f(u) = \int_0^b \frac{v}{b^2} dv$ for $0 < u \leq 1$, and $f(u) = \int_0^{b/u} v/b^2 dv = 1/(2u^2)$ for $1 < u < \infty$.

$$f(u) = \begin{cases} 1/2 & \text{for} \quad 0 < u < 1; \\ 1/(2u^2) & \text{for} \quad 1 < u < \infty \end{cases}$$

which is independent of the parameter b.

2.4   APPLICATIONS

This distribution finds applications in many fields. For instance, heat conductivity, thermal diffusion, and voltage fluctuations in a short time period (temporal dimension) are assumed to be uniform distributed. Some of these properties can also be extended to spatial dimensions (areas that are unit distance away from the source). It is used in nonparametric tests like Kolmogorov-Smirnov test. The rounding errors resulting from grouping data into classes uses a U(0,1) to obtain a correction factor known as Sheppard's correction. Quantization errors in audio-coding (compression) use this distribution. It is also used in stratified sampling, non-random clustering, etc. Random numbers from other distributions are easy to generate using U[0,1]. These are discussed in subsequent chapters.

Example 2.38   Estimating proportions   A jar contains a mixture of two liquids, L_1 and L_2, that mixes well in each other (as water and wine, or acid and water). All that is known is that "there is at most three times as much of one as the other." Find the probability that (i) $L_1/L_2 \leq 2$ and (ii) $L_1/L_2 \geq 1$.

Solution 2.39   The given condition is $\frac{1}{3} \leq L_1/L_2 \leq 3$. Let $U = L_1/L_2$. Assume that U is uniformly distributed in $[1/3, 3]$. As $3 - 1/3 = 8/3$, we take the density function as $f(x) = 3/8$, $\frac{1}{3} \leq x \leq 3$. The required answer for (i) is $P[U \leq 2] = \int_{1/3}^2 f(x)dx = (3/8) * x|_{1/3}^2 = (3/8) * (2 - 1/3) = 5/8$; and (ii) $L_1/L_2 \geq 1 = \int_1^3 f(x)dx = (3/8) * x|_1^3 = 6/8 = 0.75$.

2.5   TRIANGULAR DISTRIBUTIONS

Triangular distributions are so named because the figure looks like a triangle. The PDF is given by

$$f(x; a, b) = \begin{cases} 0 & \text{for} \quad x < a \\ 2(x-a)/[(b-a)(c-a)] & \text{for} \quad a \leq x \leq c; \\ 2(b-x)/[(b-a)(b-c)] & \text{for} \quad c \leq x \leq b; \\ 0 & \text{otherwise,} \end{cases} \tag{2.20}$$

where a is the lower and b is the upper limit, and c is the x value for the peak.

This distribution is symmetric when $c - a = b - c$, in which case the mean is $(a + b + c)/3$ and variance is $(b - a)^2/24$. A special case is the zero-centered triangular distribution for which a is negative. A triangular distribution defined over $(-a, +a)$ has PDF $f(x; a) = \frac{1}{a}(1 - |x|/a)$.

Problem 2.40   During the COVID-19 pandemic, the travel cost of a patient from home to the doctor's office is $my\$$ for being m minutes early in the office, and $mz\$$ for being m minutes late. Show that the optimal start time from home to minimize the expected cost is $z/(z + y)$.

2.6   SUMMARY

This chapter introduced the continuous uniform distributions, and its basic properties. Several statistical distributions can be obtained as nonlinear transformation of U(0,1). Thus, random numbers from such distributions are generated using uniform random numbers. A brief discussion of triangular distribution also appears in this chapter.

CHAPTER 3

Exponential Distribution

3.1 INTRODUCTION

The exponential distribution (denoted by EXP(λ)) is a continuous distribution with a single unknown parameter (usually denoted by λ, α, θ, or c), which is a positive real number. It can be regarded as the continuous analogue of geometric distribution. The PDF is given by

$$f(x; \lambda) = \begin{cases} \lambda \, \exp(-\lambda x) & \text{for} \quad x \geq 0, \lambda > 0; \\ 0 & \text{otherwise.} \end{cases}$$

This is called the one-parameter exponential distribution. The inter-arrival times in queuing systems, in website hits (time interval within which two random web users visit the website), etc., and testing product reliability uses this distribution. The unknown λ, called the *disintegration constant* in physics, and *rate* in some fields, is a scale parameter. When $\lambda = 1$, we get the standard exponential distribution (SED) EXP(1) with PDF $f(x) = e^{-x}$. Write $\exp(-\lambda) = \theta$ to get $f(x, \theta) = -\log(\theta)\theta^x = \log(1/\theta)\theta^x$. Multiply numerator and denominator by $1 - \theta$ to get $f(x, \theta) = C\theta^x(1 - \theta)$, where $C = -[\log(\theta)/(1 - \theta)]$. This reveals its resemblance with the geometric distribution. This form is also useful in extending EXP(λ) to multi-parameter versions. Both of them are applicable to statistical processes that decline monotonically. To decide whether an exponential distribution is the most appropriate one, we need to check if prior events are independent of future ones, and if the events occur at a fixed rate.

The importance of the exponential distribution in theoretical statistics arises from the classification of distributions belonging to the exponential family (or exponential class). Most of the well-known distributions like binomial distribution, Geometric distribution, Poisson distribution, negative binomial distribution, beta distribution, normal distribution, lognormal distribution, gamma distribution, inverse Gaussian distribution, χ^2 distribution, etc., belong to this family.

3.1.1 ALTERNATE REPRESENTATIONS

Setting $\lambda = 1/\theta$ gives an alternative representation as $f(x, \theta) = (1/\theta)\exp(-x/\theta)$. Next, put $\lambda = -\theta$ to get $f(x, \theta) = -\theta \exp(\theta x)$. This form is used in probabilistic degradation processes of engineering chemistry. The inverse exponential distribution discussed on page 29 results from the transformation $Y = 1/X$. A change of origin and scale transformation results in the origin

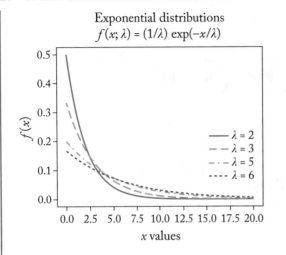

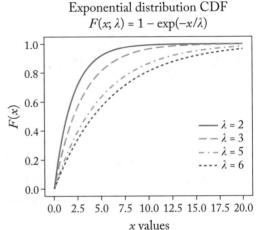

Figure 3.1: Exponential distributions.

shifted distribution

$$f(x; \lambda, a, b) = \begin{cases} (\lambda/b) \exp(-\lambda(x-a)/b), & x \ge a \\ 0 & \text{otherwise,} \end{cases} \qquad (3.1)$$

which reduces to $f(x; a, b) = (1/b)\exp(-(x-a)/b)$ for $\lambda = 1$, from which $f(x; b) = (1/b)\exp(-x/b)$ for $a = 0$, and $f(x; a) = \exp(-(x-a))$ for $b = 1$, which is the location transformed exponential distribution.

3.2   RELATION TO OTHER DISTRIBUTIONS

It is a special case of gamma distribution with $m = 1$ (Chapter 6), and Weibull distribution. The Rosin–Rammler–Bennett (RRB) distribution used in mineral engineering is a special case of one-parameter exponential distribution. If $X \sim \text{EXP}(\lambda)$, and b is a constant, then $Y = X^{1/b} \sim \text{WEIB}(\lambda, b)$. The sum of n IID exponential variates with the same parameter is gamma (also called Erlang) distributed, and with different parameters has a hyper-exponential distribution. Similarly, if $X_1, X_2, \ldots, X_n$ are IID $\text{EXP}(\lambda)$ and $S_n = X_1 + X_2 + \cdots + X_n$, then $\Pr[S_n < t < S_{n+1}]$ has a Poisson distribution with parameter λt. It is also related to the U(0,1) distribution, and power-law distribution, which is another discrete analogue of this distribution [34]. The extreme value distribution can be considered as a nonlinear generalization of $\text{EXP}(\lambda)$. Its relationship with zero-truncated Poisson (ZTP) distribution is used to generate random numbers [98]. If X_1 and X_2 are IID $\text{EXP}(\lambda)$, then $Y = X_1/(X_1 + X_2) \sim \text{U}(0,1)$. This has the implication that "if two random numbers between 0 and 1 are chosen from U(0,1), the ratio of one of them to their sum is more likely to be close to half; but when the two numbers are

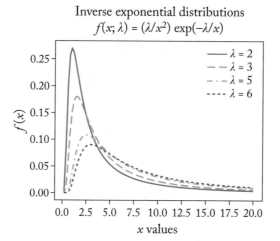

Figure 3.2: Inverse exponential distribution.

chosen from an exponential distribution, the same ratio is uniform between 0 and 1." Putting $Y = 1/X$ results in the inverse exponential distribution (Figure 3.2) with PDF

$$f(y; \lambda) = (\lambda/y^2) \exp(-\lambda/y). \tag{3.2}$$

If X_1 and X_2 are IID EXP(1) random variables, $Y = X_1 - X_2$ has standard Laplace distribution. A mixture of exponential distributions with gamma mixing weights gives rise to Lomax distribution.

3.3   PROPERTIES OF EXPONENTIAL DISTRIBUTION

This distribution has a single parameter, which is positive. It is a reverse-J shaped distribution which is always positively skewed (see Figure 3.1; page 28). Variance of this distribution is the square of the mean, as shown below. This means that when $\lambda \to 0$, the variance and kurtosis increases without limit (see Figure 3.1; page 28). The CDF is given by

$$F(x; \lambda) = \begin{cases} 1 - \exp(-\lambda x), & x \geq 0 \\ 0 & \text{otherwise.} \end{cases} \tag{3.3}$$

The SF is $S(x; \lambda) = 1\text{-CDF} = \exp(-\lambda x)$, so that $f(x; \lambda) = \lambda S(x; \lambda)$. From this the hazard function is obtained as

$$h(x) = f(x)/[1 - F(x)] = f(x)/S(x) = \lambda, \tag{3.4}$$

which is constant. When a device or an equipment approximately exhibits constant hazard rate, it is an indication that the EXP(λ) may be a good choice to model the lifetime. If x_α is the αth percentage point, $\alpha = 1 - \exp(-\lambda x_\alpha)$ from which $x_\alpha = -\log(1-\alpha)/\lambda$.

Problem 3.1   A left-truncated exponential distribution with truncation point c has PDF $f(x;\lambda) = \lambda e^{-\lambda x}/[1 - e^{-c\lambda}]$ for $x > c$. Obtain the mean and variance.

Problem 3.2   Prove that the exponential distribution is the continuous-time analog of the geometric distribution.

Solution 3.3   Let T denote the lifetime of a non-repairable item (like light-bulbs, transistors, micro-batteries used in watches, etc.) that wears out over time. Divide T into discrete time units of equal duration (say c). This duration may be counted in hours for light bulbs, days for transistors, and so on. Thus, the time-clicks are counted in unit multiples of c (say 1200 hours for light bulbs). Let N denote the number of time-clicks until the item fails, so that $T = Nc$. This equation connects the continuous lifetime with discrete time-clicks. Assume that N has a geometric distribution GEO($c\lambda$) where $c\lambda$ denotes the failure probability for each time-click. Then $\Pr[N = n] = (1 - c\lambda)^{n-1} (c\lambda)$, where $(1 - c\lambda)^{n-1}$ denotes the probability that the unit did not fail during the first $(n - 1)$ time-clicks. As the SF of T is $\Pr[T > k] = \Pr[Nc > k] = \Pr[N > \lfloor k/c \rfloor] = (1 - c\lambda)^{\lfloor k/c \rfloor}$, where the integer part is taken because N is discrete. This is of the form $(1 - \lambda/m)^m$, which as $m \to \infty$ tends to $\exp(-\lambda)$ with $m = 1/c$. As $c \to 0$, the SF approaches $\exp(-k\lambda)$, which is the SF of exponential distribution.

Problem 3.4   If $X \sim EXP(\lambda)$, find (i) $\Pr[1 \le X \le 2]$ and (ii) $\Pr[X \ge x]$ when $\lambda = \ln(2)$.

Problem 3.5   If $X \sim EXP(\lambda)$ and $Y \sim EXP(\mu)$ prove that $\Pr(X < Y) = \lambda/(\lambda + \mu)$.

This distribution represents the time for a continuous process to change state (from working to non-working, from infected to recovery, detected to non-detected or *vice versa*). For example, the time between detection of radioactivity by a Geiger counter (absence to presence) is approximately exponentially distributed.

3.3.1   MOMENTS AND GENERATING FUNCTIONS

The characteristic function is readily obtained by integration as $\phi_x(t;\lambda) =$

$$\int_0^\infty e^{itx} \lambda e^{-\lambda x} dx = \lambda \int_0^\infty e^{-(\lambda - it)x} dx = \frac{\lambda}{(\lambda - it)} = \frac{1}{1 - \frac{it}{\lambda}} = (1 - it/\lambda)^{-1}. \tag{3.5}$$

Expand as an infinite series using $(1 - x)^{-1} = 1 + x + x^2 + x^3 + \cdots$ to get

$$(1 - it/\lambda)^{-1} = 1 + it/\lambda + (it/\lambda)^2 + (it/\lambda)^3 + \cdots. \tag{3.6}$$

From this the mean and variance follow as $\mu = 1/\lambda$ and $\sigma^2 = 1/\lambda^2$. Alternately, the mean is given by $\mu = \lambda \int_0^\infty x e^{-\lambda x} dx$. Write the integral as $\int_0^\infty x^{2-1} e^{-\lambda x} dx$. Using gamma integral this becomes $\mu = \lambda \Gamma(2)/\lambda^2$. One λ cancels out and we get $\mu = 1/\lambda$ as $\Gamma(2) = 1$.

Problem 3.6   Periodic heating occurs in some electronic devices and circuits like switching, routers, and various sensors. The in-depth attenuation of surface temperature of a device surface in such cases is given by $T(x) = C * T(0) \exp(-x \sqrt{\omega/(2\alpha)})$ for $x > 0$, where ω, in radians per second, is the sinusoidal frequency, α is thermal diffusivity of the material and $T(0)$ is the initial temperature. Find the unknown C, the mean, and modal value.

For the alternate form $f(x, \theta) = \frac{1}{\theta} e^{-x/\theta}$, the mean $\mu = \theta$ and variance $\sigma^2 = \theta^2$. The general form of the exponential distribution is given by

$$f(x; c, \lambda) = (1/\lambda) e^{-(x-c)/\lambda}, \quad x \geq c, \quad \lambda > 0. \tag{3.7}$$

The corresponding CDF is $F(x; c, \lambda) = 1 - \exp(-(x - c)/\lambda)$, and characteristic function is

$$\phi(t) = \frac{e^{ict}}{1 - it/\lambda}. \tag{3.8}$$

The coefficient of skewness and kurtosis are 2 and 9, respectively. Hence, the distribution is always asymmetric, positively skewed, and leptokurtic. As it tails off slowly, it is called a "light-tailed" distribution. It is a "one-sided" distribution for small λ values. The kth moment is easily found using gamma function as $\mu_k' = E(X^k) = \Gamma(k + 1)/\lambda^k$. See Table 3.1 for further properties.

3.3.2   ADDITIVITY PROPERTY

Several statistical distributions obey the additivity property. This information is useful while modeling data from two or more identical populations. The sum of k independent exponentially distributed random variables EXP(λ) has a gamma distribution with parameters k and λ. Symbolically, if X_i are IID EXP(λ), then $\sum_{i=1}^k X_i \sim$ GAMMA(k, λ). This is easily proved using the MGF. If X_i are IID EXP(λ_i), then $\sum_{i=1}^k X_i$ has a hyper-exponential distribution. For $k = 2$, the sum $Y = X_1 + X_2$ has PDF

$$f(y; \lambda_1, \lambda_2) = [\lambda_1 \lambda_2/(\lambda_1 - \lambda_2)][\exp(-\lambda_2 y) - \exp(-\lambda_1 y)]. \tag{3.9}$$

3.3.3   MEMORY-LESS PROPERTY

Exponential distribution is the only continuous distribution that has memory-less property $\Pr(X \geq s + t)|\Pr(X \geq s) = \Pr(X \geq t)$ for $s, t \geq 0$.

Consider $\Pr(X \geq s + t) \cap \Pr(X \geq s)/\Pr(X \geq s)$. The numerator simplifies to $\Pr(X \geq s + t) = \lambda \int_{x=(s+t)}^\infty \exp(-\lambda x) dx = \exp(-\lambda(s + t))$ using $e^{-\infty} = 0$. The denom-

Table 3.1: Properties of exponential distribution $(\lambda e^{-\lambda x})$

Property	Expression	Comments
Range of X	$x \geq 0$	Continuous
Mean	$\mu = 1/\lambda$	
Median	$\log(2)/\lambda$	69.31% of mean
Variance	$\sigma^2 = 1/\lambda^2$	$\sigma^2 = \mu^2$
Skewness	$\gamma_1 = 2$	Never symmetric
Kurtosis	$\beta_2 = 9$	Always leptokurtic
Mean deviation	$E|X - \mu| = 2/(e\lambda)$	$2\mu_2 f_m$
CV	1	
CDF	$1 - e^{-\lambda x}$	SF = 1 − CDF = $e^{-\lambda x}$
Moments	$\mu'_r = 1/\lambda^r$	
MGF	$\lambda/(\lambda - t)$	
ChF	$\lambda/(\lambda - it)$	

Replace λ by $1/\lambda$ for the alternate parametrization.

inator is $\lambda \int_{x=s}^{\infty} \exp(-\lambda x)dx$. This simplifies to $\exp(-\lambda s)$. Taking the ratio of these gives $e^{-\lambda(s+t)}/e^{-\lambda s} = e^{-\lambda t}$, which is the RHS.

Problem 3.7   If $X \sim \text{EXP}(\lambda)$ with PDF $f(x; \lambda) = \lambda \exp(-\lambda x)$, find the probability that $\Pr(X - 1/\lambda) < 1$.

Problem 3.8   Prove that the $\text{EXP}(\lambda)$ is infinitely divisible.

Problem 3.9   If X and Y are IID $\text{EXP}(\lambda)$, prove that the ratio $Z = X/Y$ has PDF $f(z) = 1/(1 + z)^2$, for $z > 0$.

Problem 3.10   If $X_1, X_2, \ldots, X_n$ is a random sample of size n from IID EXP(1) (standard exponential distribution), prove that the minimum $X_{(1)}$ is distributed as $f(x) = n \exp(-nx)$, and maximum $X_{(n)}$ is distributed as $g(x) = n(1 - \exp(-x))^{n-1} \exp(-x)$.

Example 3.11   Distribution of $Y = \lfloor X \rfloor$ of an exponential distribution   If $X \sim \text{EXP}(\lambda)$, find the distribution of the integer part $Y = \lfloor X \rfloor$.

Solution 3.12   As X is continuous, $\Pr[Y = y] = \Pr[y \leq X < y + 1]$. Now consider

$$\Pr[y \leq X < y + 1] = \int_{y}^{y+1} \lambda \, \exp(-\lambda x)dx = -\exp(-\lambda x)|_y^{y+1}$$

$$= \exp(-\lambda y) - \exp(-\lambda(y + 1)) = \exp(-\lambda y)[1 - \exp(-\lambda)]. \tag{3.10}$$

Write $\exp(-\lambda y)$ as $[\exp(-\lambda)]^y$. Then (3.10) is of the form $q^y p = (1-q)q^y$ where $q = \exp(-\lambda)$. This is the PMF of a geometric distribution with probability of success $p = 1 - q = [1 - \exp(-\lambda)]$. Hence, $Y = \lfloor X \rfloor$ is GEO$([1 - \exp(-\lambda)])$. An interpretation of this result is that a geometric distribution GEO$(1 - \exp(-\lambda))$ can be obtained by discretizing an exponential distribution with mean $m = 1/\lambda$.

Example 3.13 Distribution of Fractional Part   If X has an exponential distribution, find the distribution of the fractional part $Y = X - \lfloor X \rfloor$.

Solution 3.14   It was shown above that if X has an exponential distribution, the distribution of $Y = \lfloor X \rfloor$ is GEO$(1 - \exp(-\lambda))$. The possible values of $Y = X - \lfloor X \rfloor$ are $0 \leq y \leq 1$. If we assume that the integer and fractional parts are independent, Y is the difference between an exponential and geometric random variables. This is of mixed type (as geometric distribution is discrete), where the continuous distribution dominates. This means that Y has a continuous distribution. It is easy to show that Y is distributed as $f(y) = \lambda \exp(-\lambda y)/[1 - \exp(-\lambda)]$, for $0 \leq y \leq 1$.

Example 3.15 Distribution of the minimum for EXP(λ)   If $X_1, X_2, \ldots, X_n$ are IID EXP(λ), find the distribution of $Y = X_{(1)} = \min(X_1, X_2, \ldots, X_n)$.

Solution 3.16   Let $S_1(y)$ and $F_1(y)$ denote the SF and CDF of $Y = X_{(1)}$, and $F(x)$ denote the CDF of X. Then $S_1(y) = 1 - F_1(y) = \Pr[Y > y] = \Pr[X_1 > y]\Pr[X_2 > y]\ldots\Pr[X_n > y]$ because of independence. As each of the X_i are IID, this becomes $S_1(y) = [1 - F(y)]^n$. As the CDF of $X \sim EXP(\lambda)$ is $[1 - \exp(-\lambda x)]$, we get the SF as $\exp(-\lambda x)$. Substitute in the above to get the SF as $S_1(y) = [\exp(-\lambda y)]^n = \exp(-n\lambda y)$. This is the SF of EXP($n\lambda$). Hence, $Y = \min(X_1, X_2, \ldots, X_n)$ has an exponential distribution with parameter $n\lambda$. If the X_i are independently distributed with different parameters λ_i as EXP(λ_i), then $Y = X_{(1)} \sim$ EXP(λ) where $\lambda = \lambda_1 + \lambda_2 + \cdots + \lambda_n$. An interpretation of this result is that if we have a serial system (hardware or software) where each unit works independently and has lifetime distribution EXP(λ_i), then the lifetime of the entire system is distributed as EXP($\lambda_1 + \lambda_2 + \cdots + \lambda_n$).

Example 3.17 Distribution of the maximum for EXP(λ)   If $X_1, X_2, \ldots, X_n$ are IID EXP(λ), find the distribution of $Y = X_{(n)} = \max(X_1, X_2, \ldots, X_n)$.

Solution 3.18   Let $F_1(y)$ denote the CDF of $Y = X_{(n)}$, and $F(x)$ denote the CDF of X. Then $F_1(y) = \Pr[Y \leq y] = \Pr[X_1 \leq y]\Pr[X_2 \leq y]\ldots\Pr[X_n \leq y]$ because of independence. As each of the X_i's are IID, this becomes $F_1(y) = [F(y)]^n$. Substitute for F(y) to get $F_1(y) = [1 - \exp(-\lambda y)]^n$. The PDF of Y is obtained by differentiation as $f(y) = n\lambda[1 - \exp(-\lambda y)]^{n-1}\exp(-\lambda y)$. An interpretation of this result is that if we have a parallel system (hardware or software) where each unit works independently and has lifetime distribution $EXP(\lambda_i)$, then the lifetime of the entire system has CDF $F_1(y) = \prod_{k=1}^{n}[1 - \exp(-\lambda_k y)]$.

Problem 3.19 If $X_1, X_2, \ldots, X_n$ are IID $EXP(\lambda)$, and $Y = X_{(n)} = \max(X_1, X_2, \ldots, X_n)$, prove that $E(Y) = H_n/\lambda$ where $H_n = 1 + 1/2 + 1/3 + \cdots + 1/n$, and $V(Y) = (1 + 1/2^2 + 1/3^2 + \cdots + 1/n^2)/\lambda^2$.

Problem 3.20 If a 911 emergency center receives 8 calls on the average per hour, what is the probability that (i) no calls are received in next 10 min? or (ii) at least 2 calls are received in 10 min.

Problem 3.21 If $X_1, X_2, \ldots, X_n$ are IID $EXP(\lambda)$, find the distribution of $V = \max(X_1, X_2, \ldots, X_n)$ and $U = \min(X_1, X_2, \ldots, X_n)$, and prove that $W = V - U$ and U are independent.

Problem 3.22 If $X \sim EXP(\lambda)$, find the PDF of floor(X) and ceil(X).

Problem 3.23 If $f(x; k) = ke^{-kx}$, $x \geq 0$, find the distribution of $Y = \sqrt{x}$.

Problem 3.24 If X and Y are IID exponentially distributed, find the distribution of $X + Y$ and $X - Y$.

Problem 3.25 Suppose $X_1, X_2, \ldots, X_N$ are IID $EXP(\lambda)$ where N is a 1-truncated geometric variate with PDF $f(x) = q^{x-2}p$, for $x = 2, 3, \ldots$. Find the distribution of $Y = X_1 + X_2 + \cdots + X_N$.

Problem 3.26 Suppose a system has two critical components connected in parallel with lifetime distribution $EXP(\lambda_k)$, for $k = 1, 2$. It works as long as at least one of the components works. Prove that the distribution of the lifetime of the system is $F(x) = (1 - \exp(-\lambda_1 x))(1 - \exp(-\lambda_2 x))$.

Problem 3.27 A random sample $X = (x_1, x_2, \ldots, x_n)$ is drawn from an $EXP(\lambda)$ population. Find the log-probability of $X|\lambda$. Prove that the MLE of λ is $\hat{\lambda} = 1/\bar{x}$.

Problem 3.28 If X_1, X_2, and $X_3 \sim$ IID $EXP(\lambda)$, find the distribution of $X_1 + X_2 - X_3$.

Problem 3.29 If the joint PDF of X and Y is $f(x, y) = \exp(-(x + y))$, $x, y > 0$, prove that the PDF of $z = x/y$ is $f(z) = 1/(1 + z)^2$, for $0 < z < \infty$.

Problem 3.30 Are X and Y independent if the joint PDF is $f(x, y) = x \exp(-(x + y))$, $x, y > 0$?

Example 3.31 Median of exponential distribution Find the median of exponential distribution with PDF $f(x, \lambda) = \lambda e^{-\lambda x}$.

Solution 3.32   Let M be the median. Then $\int_M^\infty \lambda e^{-\lambda x} dx = 0.50$. This gives $-e^{-\lambda x}|_M^\infty = 1/2$, or equivalently $e^{-\lambda M} = 1/2$. Take log of both sides to get $-\lambda M = -\log(2)$, or $M = -\log(1/2)/\lambda = \log(2)/\lambda$ where the log is to the base e. As $\log(2) = 0.69314718$, the median is always less than the mean (69.31% of mean).

Problem 3.33   Prove that for the exponential distribution mean/median $= \log(2)$.

Example 3.34 Quartiles of exponential distribution   Find the quartiles of $\text{EXP}(\lambda)$ with PDF $f(x, \lambda) = \lambda\, e^{-\lambda x}$.

Solution 3.35   The first quartile Q_1 is found by solving $1 - \exp(-\lambda Q_1) = 1/4$, from which $Q_1 = -\log(3/4)/\lambda \approx 0.28768/\lambda$. The third quartile Q_3 is $Q_3 = -\lambda \log(1/4) = \lambda \log(4) \approx 1.386294/\lambda$.

Example 3.36 $\Pr(X > \lambda/2), \Pr(X > 1/\lambda)$ **for EXP(λ) distribution**   Show that $\Pr[X > \lambda/2]$ of the exponential distribution is $e^{-\lambda^2/2}$. What is the $\Pr[X > 1/\lambda]$?

Solution 3.37   As the survival function (SF) is $e^{-\lambda x}$, $\Pr(X > \lambda/2)$ is easily seen to be the survival function evaluated for $x = \lambda/2$. This upon substitution becomes $e^{-\lambda^2/2}$. Putting $x = 1/\lambda$ in the SF we get $e^{-1} = 1/e$. Thus, the mean $1/\lambda$ of an exponential distribution divides the total frequency in $(1 - \frac{1}{e}):\frac{1}{e}$ ratio. This is a characteristic property of exponential distribution.

Problem 3.38   Find the area from 0 to $1/\lambda$ and from $1/\lambda$ to ∞ of the exponential distribution $\lambda e^{-\lambda x}$.

Problem 3.39   Prove that the mean of an exponential distribution divides the area in $(1 - \frac{1}{e}) : \frac{1}{e}$ ratio.

Problem 3.40   An insurance company notices that the number of claims received per day $X \sim \text{POIS}(\lambda)$ and claimed amounts $Y \sim \text{EXP}(10\lambda)$. What is the expected total claimed amount per day?

3.4   RANDOM NUMBERS

The easiest way to generate random numbers is using the inverse CDF method. As the quantiles of the distribution are given by $u = F(x) = 1 - \exp(-\lambda x)$, we get $x = F^{-1}(u) = -\log(1 - u)/\lambda$. This becomes $-\theta \log(1 - u)$ for the alternate form $f(x, \theta) = (1/\theta) \exp(-x/\theta)$. As $U(0, 1)$

and $1 - U$ are identically distributed (page 20), we could generate a random number in $[0,1)$ and obtain $u = -\log(u)/\lambda$ as the required random number (Marsaglia (1961) [98]).

Problem 3.41   If X, Y are IID EXP(1/2), prove that $Z = (X - Y)/2$ is Laplace distributed.

Problem 3.42   Find k for the PDF $f(x) = kx^{-p}\exp(-c/x)$, $0 < x < \infty, c > 0, p > 1$. Show that the rth moment is $E(x^r) = c^r\Gamma(p - r + 1)/\Gamma(p - 1)$ for $r \leq (p + 1)$.

Example 3.43   Mean deviation of exponential distribution   Find the mean deviation of the exponential distribution $f(x,\lambda) = \lambda e^{-\lambda x}$.

Solution 3.44   We know that the CDF is $1 - e^{-\lambda x}$. Thus, the MD is given by

$$\text{MD} = 2\int_0^{1/\lambda}(1 - e^{-\lambda x})dx. \tag{3.11}$$

Split this into two integrals and evaluate each to get

$$\text{MD} = 2[1/\lambda + (1/\lambda)\exp(-1) - (1/\lambda)] = 2/(e\lambda) = 2\mu_2 * f_m, \tag{3.12}$$

where $f_m = \lambda e^{-1} = \lambda/e$. Alternatively, use the SF() version as the exponential distribution tails off to the upper limit.

3.5   FITTING

The MLE and MoM estimators of λ are the same, and are the reciprocal of the mean. If $S = (x_1, x_2, \ldots, x_n)$ is a random sample of size n from an EXP(λ) distribution, the MLE of λ is given by $\hat{\lambda} = 1/\overline{x}_n$, which is biased. For the alternate version $f(x, \theta) = (1/\theta)\exp(-x/\theta)$, the MLE is $\hat{\theta} = \overline{x}_n$. An unbiased estimator with minimum variance is $(n - 1)/\sum_k x_k$. This estimate gets shifted by c units to the left for left-truncated exponential distribution at $x = c$. In other words, $\hat{\lambda} = 1/(\overline{x}_n - c)$.

3.6   GENERALIZATIONS

Several generalizations of the exponential distribution exist. A majority of them are multi-parameter extensions. The exponentiated exponential distribution (EED) has CDF $G(x; \lambda, k) = (1 - \exp(-\lambda x))^k$, where $k > 0$ is a real constant. Differentiate w.r.t. x to get the PDF as $g(x; \lambda, k) = k\lambda\exp(-\lambda x)(1 - \exp(-\lambda x))^{k-1}$. The size-biased exponential distribution is $g(y; k, \lambda) = C(1 + ky)\exp(-\lambda y)$ where $C = \lambda^2/(k + \lambda)$, which is a convolution of an exponential and gamma laws. See Shanmugam (1991) [123] for a test for size-bias that has applications in econometrics and finance. Higher-order weighted exponential distributions can be obtained as

$$g(y; k_1, k_2, \ldots, k_m, \lambda) = C(1 + k_1 y + k_2 y^2 + \cdots + k_m y^m)\exp(-\lambda y), \tag{3.13}$$

which are convolutions of gamma laws (where k'_js could be functions of λ). These find applications in real-life models involving exceedances (like dam capacity, rainfall, agricultural harvests, electrical voltages, atmospheric pollution, etc.). Amburn et al. (2015) [9] used such models for point probabilistic quantitative precipitation forecasts for rainfall exceedance as

$$f(x; b) = (1 + x/b + x^2/2b^2) \exp(-x/b), \tag{3.14}$$

where μ is the mean conditional quantitative precipitation forecast (cQPF), $b = \mu/\alpha$ ($\alpha = 3$ for 100% coverage, and 1 otherwise is the parameter of Gamma, and x is the exceedance threshold. Azzalini's skew exponential distribution has PDF $g(x, c) = K f(x)F(cx)$ where $c > 0$, $F()$ denotes the CDF, and K is the normalizing constant. Letting $f(x) = \exp(-x)$, we get the skew SED PDF as $g(x, c) = (1 + 1/c) \exp(-x)[1 - \exp(-cx)])$.

Problem 3.45   Let X_i's be IID EXP(λ) with PDF $f(x_i) = \frac{1}{\lambda}e^{-x_i/\lambda}$. Define new variates Y_i's as $Y_1 = X_1/(X_1 + X_2 + \cdots + X_n)$, $Y_2 = (X_1 + X_2)/(X_1 + X_2 + \cdots + X_n)$, etc., $Y_k = (X_1 + X_2 + \cdots + X_k)/(X_1 + X_2 + \cdots + X_n)$, and $Y_n = (X_1 + X_2 + \cdots + X_n)$. Prove that the joint distribution of $(Y_1, Y_2, \ldots, Y_n)$ depends upon y_n and y_{n-1} only.

Problem 3.46   Use $E|X - a| = E|X - \mu| + 2 \int_\mu^a (a - x) f(x) dx$ to find the MD from the median of an exponential distribution.

3.7   APPLICATIONS

This distribution finds applications in modeling random proportions, radioactive decay, particle diffusion, and lifetime of devices and structures.[1] The random proportion of vibrational levels (fraction of diatomic molecules in excited vibrational state) in statistical mechanics is approximately exponentially distributed, for which the shape parameter λ depends on the type of gas and the temperature. It is used to model lifetimes in reliability theory, and waiting times in queuing theory, length of calls in telephony, duration of parking time in parking lots, headway (time interval) between adjacent vehicles when traffic volume is low in highway engineering, etc., all of which are temporal processes. It can also model spacial events like distance between outbreak of infectious diseases like coronavirus in an enclosed region, distance between seedlings in a field, etc. The implied meaning is that we are more likely to see a small x-value (e.g., short time durations in queuing theory and telephony) and less likely to see large x-values.

It is the widely used distribution to model voltage drops across diodes, the time to failure (TTF) of electronic devices and equipments that fail at a constant rate, regardless of the accumulated age. The failure rate of continuous operational systems are estimated as $\lambda = r/T$ where r is the total number of failures occurring during an investigation period, T is the total running time, cycles, or miles during an investigation period for both failed and non-failed (healthy) items. The λ is the failure rate (called hazard rate in some fields). The the mean time to failure

[1]Other choices for lifetime distribution models are Weibull model, and inverted exponential distribution.

(MTTF) is the time between the first use to the time it fails. A related term is the mean time between failure (MTBF) of repairable systems. These can be modeled using the exponential law where $\lambda = 1/MTTF$.

The one-parameter exponential distribution provides only an approximate fit for some of the lifetime problems, in which case the multi-parameter version becomes useful. The time between two successive events in a Poisson process, distance between mutations in a DNA strand etc. are assumed to follow this law. Note that the unit (for time, distance, etc.) is application specific. For example, the expected life length of a new light bulb (in hours) can be assumed to follow an exponential distribution with parameter $\lambda = 1/500$ so that the lifetime is given by $f(x) = (1/500)(e^{-x/500})$. See Shanmugam (2013) [126] for a tweaked exponential distribution to estimate survival time of cancer patients, Shanmugam, Bartolucci, and Singh (2001) [130, 131] for applications in neurology, and [124] for testing guaranteed exponentiality.

Its applications in chemical processes include probabilistic degradation analysis of organic compounds and gases, corrosion analysis, etc., that are modeled as a general exponential process $X = A \exp(B(t - t_0))$, where A and B are suitable real constants, X represents final degradation at epoch t from a fixed origin t_0. This becomes the classical EXP(λ) when $A = -B = \lambda$. It is an exponential process for $B > 0$. Taking natural logarithm results in a linear model, which can be fit using least squares principle. Packaged food-processing industries use thermal treatment to pasteurize (remove micro-organisms in) food packets. Several mixing equipments like tubular equipment with plug flow or laminar flow, perfect mixing agitated vessel (PMAV), etc., are employed for this purpose. Distribution of ages of micro-organisms that leave the equipment in PMAV models is $(1/t_m) \exp(-t/t_m)$ where t_m is the mean residence time, and t denotes the time needed to remove certain percentage of the organisms.

The demand patterns of some goods and services (like medicines, essential supplies) during emergencies and epidemics are also modeled using exponential growth. Although they are unrestricted theoretically, several practical applications exhibit a curtailment or altered growth rate due to space and resource constraints. For example, tumors that grow inside the body of living organisms are limited in growth (size) due to body shape, nutrient and oxygen availability, healthy organs in its path, etc. The growth can be quantified using the doubling period (time for it to double in size).

Other examples include modeling lifetime of

- destructive devices that are (more or less) continuously or regularly in use, like light bulbs, tubes, electronic chips, etc.;

- nondestructive or reusable devices until next repair work, electronic devices like computer monitors and LCD screens, microwaves, electrical appliances like refrigerators, lifetime of automobile tires; and

- components within devices like components in automobiles or aircraft, parts like transistors and capacitors in electronic devices.

Similarly, the time until the arrival of the next event (like next telephone call, emergency call, etc.) or time until next customer to an office or business can be modeled using this distribution. The lifetime distribution of a system-in-series comprising of m components, each of which is distributed as $EXP(\lambda_k)$, is itself $EXP(\sum_{k=1}^{m} \lambda_k)$. The system-level hazard for it is the sum of the hazards of the components.

Example 3.47 Lifetime of components   The lifetime of a component is known to be exponentially distributed with mean $\lambda = 320$ hours. Find the probability that the component has failed in 340 hours, if it is known that it was in good working condition when time of operation was 325 hours.

Solution 3.48   Let X denote the lifetime. Then $X \sim EXP(1/320)$. Symbolically, this problem can be stated as $\Pr[X < 340 | X > 325]$. Using conditional probability, this is equivalent to $P[325 < X < 340]/P[X > 325]$. In terms of the PDF this becomes $\int_{325}^{340} f(x)dx / \int_{325}^{\infty} f(x)dx$. Write the numerator as $\int_{325}^{\infty} f(x)dx - \int_{340}^{\infty} f(x)dx$ so as to get $1 - e^{-340/320}/e^{-325/320} = 1 - e^{-15/320} = 0.04579$.

3.7.1   HYDRAULICS

Load-resistance interference is a technique used in hydraulics and mechanical engineering. It is called stress-strength analysis in structural engineering. The stress (or loading) of an equipment (or a structure) beyond a threshold could lead to its failure. Examples of such loadings are temperature and voltage fluctuations, flow rate (of fluids, liquids or gases), mechanical loads, high rotational rates, vibrations, etc. Mathematically, it is denoted as $\Pr[r - l > 0]$ where r is the resistance (strength) and l is the loading (stress) at any instant in time. Closed-form expressions for reliability can be obtained when both the loading and resistance are exponentially distributed. If loading $\sim EXP(\lambda)$ and resistance $\sim EXP(\mu)$, the reliability is $\lambda/(\lambda + \mu)$. The resistance in some applications follow the Gaussian or skewed Gaussian law. The reliability in these cases can be approximated using the CDF of Gaussian distribution $\Phi()$ if all the parameters are known.

3.7.2   TENSILE STRENGTHS

Strength of materials are employed by quality control specialists and test engineers for several manufactured items like threads, wires, coils, cable, etc. Denote the probability that a given length t of a product or material can sustain a fixed load by $F(t)$. Two adjacent portions of the material of length $t + s$ sustain iff the two segments individually sustain the applied load. This can be expressed as $F(t + s) = F(t)F(s)$ assuming independence of the segments. Then the length at which the material will break follows an exponential distribution.

3.8   SUMMARY

The exponential distribution used to be one of the most popular for lifetime modeling applications. Several new statistical distributions like Birnbaum-Saunders distribution, extended Weibull distribution and so on are increasingly being used for this purpose. Nevertheless, due to its simplicity and ease of fitting, this distribution is still used extensively in some fields.

CHAPTER 4

Beta Distribution

4.1 INTRODUCTION

The beta distribution has a long history that can be traced back to the year 1676 in a letter from Issac Newton to Henry Oldenbeg (see Dutka (1981) [55]). It is widely used in civil, geotechnical, earthquake, and metallurgical engineering due to its close relationship with other continuous distributions. The PDF of Beta-I(a, b) is given by[1]

$$f(x; a, b) = x^{a-1}(1 - x)^{b-1}/B(a, b),\tag{4.1}$$

where $0 < x < 1$, and $B(a, b)$ is the complete beta function (CBF) defined as $B(a, b) = \int_0^1 x^{a-1}(1 - x)^{b-1}dx$. It is called the standard or type-I beta or Beta-I distribution. Particular values for $a > 0$ and $b > 0$ results in a variety of distributional shapes. The Beta-I distribution is a proper choice in risk-modeling because the risks in many applications can be lower and upper bounded, and scaled to any desired range (say (0,1) range) [78]. Events constrained to happen within a finite interval can be modeled due to the wide variety of shapes assumed by this distribution.

It is also used in Bayesian models with unknown probabilities, in order-statistics and reliability analysis. In Bayesian analysis, the prior distribution is assumed to be Beta-I for binomial proportions. It is used to model the proportion of fat (by weight) in processed or canned food, percentage of impurities in some manufactured products like food items, cosmetics, laboratory chemicals, etc. Data in the form of proportions arise in many applied fields like marketing, toxicology, bioinformatics, genomics, etc. Beta distribution is the preferred choice when these quantities exhibit extra variation than expected. Important distributions belonging to the beta family are discussed below. These include type I and type-II beta distributions. We will use the respective notations Beta-I(a, b), and Beta-II(a, b).[2] Beta distributions with three or more parameters are also briefly mentioned.

4.1.1 ALTERNATE REPRESENTATIONS

Write $c = a - 1$ and $d = b - 1$ to get the alternate form

$$f(x; c, d) = x^c(1 - x)^d/B(c + 1, d + 1).\tag{4.2}$$

---

[1]The variable is chosen as "e" in astronomy (like exoplanet eccentricity modeling), as "c" in optics.
[2]p and q are used in place of a and b in reliability engineering.

Put $x = \sin^2(\theta)$ in (4.2) to get

$$f(\theta; c, d) = \sin^{2c}(\theta)\cos^{2d}(\theta)/B(c+1, d+1) \quad \text{for} \quad 0 < \theta < \pi/2. \tag{4.3}$$

Some applications use a and $n - a + 1$ as parameters resulting in

$$f(x; a, b, n) = x^{a-1}(1-x)^{n-a}/B(a, n-a+1). \tag{4.4}$$

A symmetric beta distribution results when $a = b$ with PDF

$$f(x; a) = x^{a-1}(1-x)^{a-1}/B(a, a) = [x(1-x)]^{a-1}\Gamma(2a)/[\Gamma(a)]^2. \tag{4.5}$$

Beta distributions defined on $(-1, +1)$ are encountered in some applications. Using the transformation $Y = 2X - 1$ we get $f(y) = f(x)/2 = f((y+1)/2)/2$. This results in the PDF

$$f(y; a, b) = [(y+1)/2]^{a-1}[(1-y)/2]^{b-1}/[2B(a, b)]. \tag{4.6}$$

This simplifies to

$$f(y; a, b) = C(1+y)^{a-1}(1-y)]^{b-1} \quad \text{where} \quad -1 < y < 1, \quad \text{and} \quad C = 1/[2^{a+b-1}B(a, b)]. \tag{4.7}$$

This also can be generalized to 4-parameters as

$$f(x; a, b, c, d) = C(1 + x/c)^{a-1}(1 - x/d)]^{b-1}, \tag{4.8}$$

and to the 6-parameters as

$$f(x; a, b, c, d, p, q) = C(1 + (x-p)/c)^{a-1}(1 - (x-q)/d)]^{b-1}, \tag{4.9}$$

where C is the normalizing constant, which is found using the well-known integral

$$\int_a^b (x-a)^{a-1}(b-x)^{b-1}dx = (b-a)^{a+b-1}B(a, b). \tag{4.10}$$

These are related to the Berstein-type basis functions $Y_k^n(x; a, b, m) = \binom{m}{k}(x-a)^k(b-x)^{n-k}/(b-a)^m$ [147]. Truncated and size-biased versions of them are used in several engineering fields.

4.2   RELATION TO OTHER DISTRIBUTIONS

It is a special case of gamma distribution with $m = 1$. It reduces to uniform (rectangular) distribution U(0,1) for $a = b = 1$. A triangular-shaped distribution results for $a = 1$ and $b = 2$, or vice versa. When $a = b = 1/2$, this distribution reduces to the arcsine distribution of first kind (Chapter 5). If $b = 1$ and $a \neq 1$, it reduces to power-series distribution $f(x; a) = ax^{a-1}$ using the result $\Gamma(a+1) = a * \Gamma(a)$. A J-shaped distribution is obtained when a or b is less than one. If X and Y are IID gamma distributed with parameters a and b, the ratio $Z = X/(X + Y)$ is Beta-I(a, b) distributed. If X'_ks are IID Beta-I$(\frac{2k-1}{2n}, \frac{1}{2n})$ random variables, the distribution of the geometric mean (GM) of them $Y = (\prod_{k=1}^{n} X_k)^{1/n}$ is SASD-I distributed (Chapter 5) for $n \geq 2$ ([90], [22]). This has the interpretation that the GM of Beta-I$(\frac{2k-1}{2n}, \frac{1}{2n})$ random variables converges to arcsine law whereas the AM tends to the normal law (central limit theorem). As χ^2 distribution is a special case of gamma distribution, a similar result follows as $Z = \chi_m^2/(\chi_m^2 + \chi_n^2) \sim$ Beta-I$(m/2, n/2)$.

As $(\chi_m^2 + \chi_n^2)$ is independent of Z, the above result can be generalized as follows: If $X_1, X_2, \ldots X_n$ are IID normal variates with zero means and variance σ_k^2, then $Z_1 = X_1^2/(X_1^2 + X_2^2)$, $Z_2 = (X_1^2 + X_2^2)/(X_1^2 + X_2^2 + X_3^2)$, and so on are mutually independent beta random variables. If X has an $F(m, n)$ distribution, then $Y = (m/n)X/[1 + (m/n)X]$ is beta distributed. The beta distribution is also related to the Student's t distribution under the transformation $x = 1/(1 + t^2/n)$. Similarly, $y = -\log(x)$ has PDF

$$f(y; a, b) = \exp(-ay)(1 - \exp(-y))^{b-1}, \tag{4.11}$$

and $y = x/(1 - x)$ results in beta-prime distribution (page 48). The positive eigenvalue of Roy's $\theta_{\max}$-criterion used in MANOVA has a Beta-I distribution when $s = \max(p, n_h) = 1$, where p is the dimensionality and n_h is the DoF of the hypothesis.

Problem 4.1   Prove that $B(a + 1, b) = [a/(a + b)]B(a, b)$ where $B(a, b)$ denotes the CBF. What is the value of $B(.5, .5)$?

Problem 4.2   If $X \sim$ Beta-I(a, b), find the distribution of $Y = (1 - X)/X$, and obtain its mean and variance. Find the ordinary moments.

Problem 4.3   If X and Y are independent gamma random variables GAMMA(a, λ) and GAMMA(b, λ), then prove that $X/(X + Y)$ is Beta(a, b).

Problem 4.4   Verify whether $f(x; c, d) = (1 + x)^{c-1}(1 - x)^{d-1}/[2^{c+d-1}B(c, d)]$ is a PDF for $-1 < x < 1$, where $B(c, d)$ is the complete beta function.

4.3   PROPERTIES OF TYPE-I BETA DISTRIBUTION

This distribution has two parameters, both of which are positive real numbers. The range of x is between 0 and 1. The variance is always bounded, irrespective of the parameter values. Put $y = 1 - x$ in the above to get the well-known symmetry relationship $f_x(a, b) = f_y(b, a)$ or in terms of tail areas $I_x(a, b) = 1 - I_{1-x}(b, a)$ where $I_x(a, b)$ is described below. If $a = b$, the distribution is symmetric about $X = 1/2$.

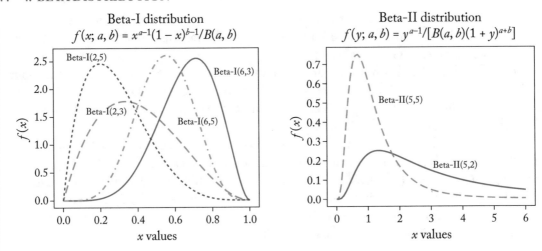

Figure 4.1: Beta distributions.

4.3.1 MOMENTS AND GENERATING FUNCTIONS OF TYPE-I BETA

The moments are easy to find using beta integral. The kth moment is

$$\mu'_k = \frac{1}{B(a,b)} \int_0^1 x^{a+k-1}(1-x)^{b-1}dx = B(a+k,b)/B(a,b) = \frac{\Gamma(a+b)\Gamma(a+k)}{\Gamma(a+b+k)\Gamma(a)}. \quad (4.12)$$

In terms of rising factorials this becomes

$$\mu'_k = a^{[k]}/(a+b)^{[k]}. \quad (4.13)$$

The mean is obtained by putting $k = 1$ as $\mu = a/(a+b) = 1 - b/(a+b)$. This has the interpretation that increasing the parameter "a" by keeping "b" fixed moves the mean to the right (toward 1). Put $k = 2$ to get the second moment as $a(a+1)/[(a+b)(a+b+1)]$. The variance is $\sigma^2 = ab/[(a+b)^2(a+b+1)]$. This is symmetric in the parameters, and increasing both "a" and "b" together decreases the variance (see Figure 4.1). If $a > 1$ and $b > 1$, there exists a single mode at $(a-1)/(a+b-2)$. The characteristic function is

$$\phi(t) = \frac{1}{B(a,b)} \int_0^1 e^{itx} x^{a-1}(1-x)^{b-1}dx = {}_1F_1(a,a+b;it), \quad (4.14)$$

where ${}_1F_1(a,a+b;it)$ is the confluent hypergeometric function. The kth central moment can be obtained as follows:

$$\mu_k = \frac{1}{B(a,b)} \int_0^1 (x - a/(a+b))^k x^{a-1}(1-x)^{b-1}dx$$
$$= (-a/(a+b))^k {}_2F_1(-k,a,a+b,(a+b)/a), \quad (4.15)$$

Table 4.1: Properties of Beta-I distribution

Property	Expression	Comments
Range	$0 \leq x \leq 1$	Continuous
Mean	$\mu = a/(a + b)$	$= 1 - b/(a + b)$
Variance	$ab/[(a + b)^2(a + b + 1)] = \mu(1 - \mu)/(a + b + 1)$	$\Rightarrow \mu > \sigma^2$
Mode	$(a - 1)/(a + b - 2)$	$a > 1, b > 1$
CV	$(b/[a(c + 1)])^{1/2}$	$c = a + b$
Skewness	$\gamma_1 = 2(b - a)\sqrt{a + b + 1}/[\sqrt{ab}(a + b + 2)]$	$a = b \Rightarrow$ symmetry
Kurtosis	$\beta_2 = 3c(c + 1)(a + 1)(2b - a)/[ab(c + 2)(c + 3)]$	$c = a + b$
Mean deviation	$E\|X - \mu\| = 2a^a b^b/[B(a, b)(a + b)^{a+b+1}]$	$2c[I_c(a, b) - I_c(a + 1, b)],$ $c = a/(a + b)$
Moments	$\mu'_r = \prod_{i=0}^{r-1}(a + i)/(a + b + i)$	$a^{(r)}/(a + b)^{(r)}$
Moments	$\mu_r = (-c)^r {}_2F_1(a, -r, a + b, 1/c)$	$c = a/(a + b)$
ChF	$\frac{\Gamma(a+b)}{\Gamma(a)} \sum_{j=0}^{\infty} \frac{\Gamma(a+j)(it)^j}{\Gamma(a+b+j)\Gamma(1+j)}$	$_1F_1(a, a + b; it)$
Additivity	$\sum_{i=1}^{m} Beta(a_i, b) = Beta(\sum_{i=1}^{m} a_i, b)$	Independent
Recurrence	$f(x; a + 1, b)/f(x; a, b) = (1 + b/a)x$ $f(x; a, b + 1)/f(x; a, b) = (1 + a/b)(1 - x)$	
U-shaped	$a < 1$ and $b < 1$	
J-shaped	$(a - 1) * (b - 1) < 0$	
Tail area	$I_x(a, b) = \frac{1}{B(a, b)} \int_0^x t^{a-1}(1 - t)^{b-1} dt$	I = incomplete beta

Symmetric when $a = b$. The $_2F_1()$ is hypergeometric function which is related to the incomplete beta function as $I_x(a, b) = [f(x; a + 1, b)/(a + b)] * {}_2F_1(1 - b, 1; a + 1; -x/(1 - x))$ where $f(x; a + 1, b)$ is the density of Beta-I. It can also be represented using the Euler identity $_2F_1(a, -r, a + b, 1/c) = (1 - 1/c)^{b+r} {}_2F_1(b, a + b + r; a + b; 1/c)$.

where $_2F_1(a, b, c; x)$ is the hypergeometric function. The coefficient of skewness is $\gamma_1 = 2(b - a)\sqrt{a + b + 1}/[\sqrt{ab}(a + b + 2)]$. It can accommodate both positive $(a < b)$ and negative $(a > b)$ skewness. See Table 4.1 for further properties.

Problem 4.5 For a beta distribution with mean μ and standard deviation σ, prove that the shape parameters are given by $a = \mu D$ and $b = (1 - \mu)D$ where $D = (\mu'/\sigma^2 - 1)$, and $\mu' = \mu(1 - \mu)$.

Problem 4.6   If X is Beta-I(p,q), find the distribution of $Y = (1-X)/X$.

Problem 4.7   If X is Beta-I(p,q) distributed, show that the variate $Y = -\ln(X/(1-X))$ is generalized logistic(p,q) distributed.

Problem 4.8   If X is Beta-I(p,q) distributed, prove that $E(1/(1-x)) = (p+q-1)/(q-1)$.

Example 4.9   Mean vs. variance of Beta-I   Prove that the variance can never equal the mean of a beta-I distribution.

Solution 4.10   We know that the variance of Beta-I can be represented in terms of the mean as $\mu(1-\mu)/(a+b+1)$. Assume the contrary that the variance can be equal to the mean. Put $\mu = x$ in the above to get $x(1-x)/(a+b+1) = x$. This simplifies to $-x^2 = (a+b)x$. As the mean cannot be zero (as "a" cannot be zero) there is no solution possible. Hence, the variance of Beta-I is always less than the mean. Alternatively, divide the variance by the mean and argue that as $\mu \in (0,1)$, $1-\mu$ is always less than 1, showing that the ratio is < 1, which implies that $\sigma^2 < \mu$.

The inverse moments of Beta-I distribution can be obtained either using the property that the ratio $X/(X+Y)$ is Beta-I distributed when X and Y are IID gamma distributed; or using the density recurrence relationships (Chattamvelli and Jones (1995) [38]). Let X and Y be IID gamma distributed random variables with parameters a and b. Consider the ratio $Z = X/(X+Y)$. Then $Z \sim$ Beta-I(a,b). As X and Y are IID, $X+Y \sim$ Gamma$(a+b)$. It is shown in Chapter 6 that $E(X^k) = \Gamma(a+k)/\Gamma(a)$. Hence, $E(Z^k) = [\Gamma(a+k)/\Gamma(a)]/[\Gamma(a+b+k)/\Gamma(a+b)]$.

Problem 4.11   If X_k are IID unit rectangular variates for $k = 1,2,\ldots n$ with PDF $f(x) = 1$, for $0 \le x \le 1$, and the corresponding order statistics are $X_{(1)} \le X_{(2)} \le \ldots \le X_{(n)}$, prove that the rth order statistic $X_{(r)} \sim$ Beta-I$(r, n-r+1)$.

Example 4.12   Mean deviation of beta distribution   Prove that the mean deviation about the mean is given by

$$E|X - \mu| = 2a^a b^b/[B(a,b)(a+b)^{a+b+1}]. \tag{4.16}$$

Solution 4.13   The MD $= 2\int_0^c I_x(a,b)dx$ where $c = a/(a+b)$ is the mean. Taking $u = I_x(a,b)$ and $dv = dx$ this becomes

$$\text{MD} = 2\left[xI_x(a,b)|_0^c - \int_0^c xg_x(a,b)dx \right] = 2\left[c * I_c(a,b) - \int_0^c xg_x(a,b)dx \right]. \tag{4.17}$$

Write $x g_x(a,b)$ as $x^a(1-x)^{b-1}/B(a,b)$. Multiply numerator and denominator by $B(a+1,b)$ and write $B(a+1,b)/B(a,b)$ as $a/(a+b)$ to get $2\int_0^c x * g_x(a,b)dx = 2a/(a+b)I_c(a+1,b)$. This gives

$$\text{MD} = 2c\left[I_c(a,b) - I_c(a+1,b)\right], \quad \text{where} \quad c = a/(a+b). \tag{4.18}$$

This can be simplified using a result in Chattamvelli (1995) [31] as

$$I_c(a,b) - I_c(a+1,b) = (1-c)^2/(b-1)\, g_c(a+1,b-1), \tag{4.19}$$

where g( ) is the PDF of Beta-I. This gives $\text{MD} = 2c(1-c)^2/(b-1)\, g_c(a+1,b-1)$. Substitute for c and simplify to get the above form above. Alternately, use

$$I_c(a,b) - I_c(a+1,b) = c^a(1-c)^b/[a\, B(a,b)] \tag{4.20}$$

to get $\text{MD} = 2bc/[(a+b)(a+b+1)]g_c(x;a,b)$, where $c = a/(a+b)$.

4.4 TYPE-II BETA DISTRIBUTION

Beta distribution of the second kind (also called type-II beta distribution, beta-prime distribution, or inverted beta distribution (IBD)) is obtained from the above by the transformation $Y = X/(1-X)$ or equivalently $X = Y/(1+Y)$. When $x \to 0$, $y \to 0$, and when $x \to 1$, $y \to \infty$. Hence, the range of Y is from 0 to ∞. The PDF is given by

$$f(y;a,b) = y^{a-1}/[B(a,b)(1+y)^{a+b}], \quad y > 0, \ a,b > 0. \tag{4.21}$$

The Beta-I distribution is used to model random experiments or occurrences that vary between two finite limits, that are mapped to the $(0,1)$ range, while Beta-II is used when upper limit is infinite. It is also used in risk analysis in finance and marketing, etc.

4.4.1 RELATION TO OTHER DISTRIBUTIONS

Put $a = b = 1$ to get Beta$(1,1)$, which is identical to U(0,1) distribution. If X is Beta-I(a,b) then $(1-X)/X$ is Beta-II(b,a), and $X/(1-X)$ is Beta-II(a,b). If X and Y are independent gamma random variables GAMMA(a,λ) and GAMMA(b,λ), then $X/(X+Y)$ is Beta(a,b). As gamma and χ^2 are related, this result can also be stated in terms of normal variates as follows. If X and Y are independent normal variates, then $Z = X^2/(X^2+Y^2)$ is Beta-I distributed. In addition, if $X_1, X_2, \ldots, X_k$ are IID N(0,1) and $Z_1 = X_1^2/(X_1^2+X_2^2)$, $Z_2 = (X_1^2+X_2^2)/(X_1^2+X_2^2+X_3^2)$, and so on, $Z_j = \sum_{i=1}^{j} X_i^2 / \sum_{i=1}^{j+1} X_i^2$, then each of them are Beta-I distributed, as also the product of any consecutive set of Z_j's are beta distributed. The logistic distribution and type II beta distribution are related as $Y = -\ln(X)$. If X is Beta-I(a,b) then $Y = \ln(X/(1-X))$ has a generalized logistic distribution. Dirichlet distribution is a generalization of beta distribution. Order statistic from uniform distribution is beta distributed. In general, j^{th} highest order statistic from a uniform distribution is Beta-I$(j, n-j+1)$.

4.5 PROPERTIES OF TYPE-II BETA DISTRIBUTION

This is a special case of the unscaled F distribution (distribution of χ_m^2/χ_n^2), or an F with the same degrees of freedom. In other words, put $Y = (m/n) * X$ in F distribution to get Beta-II distribution. If Y is Beta-II(a,b) then $1/Y$ is Beta-II(b, a). This means that $1/[X/(1-X)] = (1-X)/X$ is also beta distributed (see below). If X, Y have the inverted Dirichlet distribution, the ratio $X/(X + Y)$ has the beta distribution.

4.5.1 MOMENTS AND GENERATING FUNCTIONS OF TYPE-II BETA

The mean and variance are $\mu = a/(b-1)$ and $\sigma^2 = a(a+b-1)/[(b-1)^2(b-2)]$ for $b > 2$. Consider $E(Y^k)$

$$\int_0^\infty y^k f_y(a,b)dy = \int_0^\infty y^{a+k-1}/[B(a,b)(1+y)^{a+b}]dy. \tag{4.22}$$

Put $x = y/(1+y)$ so that $y = x/(1-x)$, $(1+y) = 1/(1-x)$ and $dy/dx = [(1-x) - x(-1)]/(1-x)^2$. This simplifies to $1/(1-x)^2$. The range of X is $[0, 1]$. Hence, (4.22) becomes

$$(1/B(a,b)) \int_0^\infty y^{a+k-1}/(1+y)^{a+b}dy = (1/B(a,b)) \int_0^1 x^{a+k-1}(1-x)^{b-k-1}dx. \tag{4.23}$$

This is $B(a+k, b-k)/B(a,b)$. Put $k = 1$ to get the mean as $\Gamma(a+1)\Gamma(b-1)\Gamma(a+b)/[\Gamma(a)\Gamma(b)\Gamma(a+b)]$. Write $\Gamma(a+1) = a\Gamma(a)$ in the numerator, and $\Gamma(b) = (b-1)\Gamma(b-1)$ in the denominator and cancel out common factors to get $\mu = a/(b-1)$. Put $k = 2$ to get the second moment as $B(a+2, b-2)/B(a,b) = \Gamma(a+2)\Gamma(b-2)\Gamma(a+b)/[\Gamma(a)\Gamma(b)\Gamma(a+b)] = a(a+1)/[(b-1)(b-2)]$. From this the variance is obtained as $a(a+1)/[(b-1)(b-2)] - a^2/(b-1)^2$. Take $\mu = a/(b-1)$ as a common factor. This can now be written as $\mu(\frac{a+1}{b-2} - \mu)$. Substitute for μ inside the bracket and take $(b-1)(b-2)$ as common denominator. The numerator simplifies to $b - a + 2a - 1 = (a+b-1)$. Hence, the variance becomes $\sigma^2 = a(a+b-1)/[(b-1)^2(b-2)]$. As $(a+1)/(b-2) - \mu = (a+b)/[(b-1)(b-2)]$, this expression is valid for $b > 2$. Unlike the Beta-I distribution whose variance is always bounded, the variance of Beta-II can be increased arbitrarily by keeping b constant (say near 2^+) and letting $a \to \infty$. It can also be decreased arbitrarily when $(a+1)/(b-2)$ tends to $\mu = a/(b-1)$. The expectation of $[X/(1-X)]^k$ is easy to compute in terms of complete gamma function as $E[X/(1-X)]^k = \frac{\Gamma(a+k)\Gamma(b-k)}{\Gamma(a)\Gamma(b)}$. See Table 4.2 for further properties.

Example 4.14 The mode of Beta-II distribution Prove that the mode of Beta-II distribution is $(a-1)/(b+1)$.

Solution 4.15 Differentiate the PDF (without constant multiplier) w.r.t. y to get

$$f'(y) = [(1+y)^{a+b}(a-1)y^{a-2} - y^{a-1}(a+b)(1+y)^{a+b-1}]/(1+y)^{2(a+b)}. \tag{4.24}$$

Table 4.2: Properties of Beta-II distribution

Property	Expression	Comments
Range of X	$0 \leq x \leq \infty$	Continuous
Mean	$\mu = a/(b-1)$	
Variance	$\sigma^2 = a(a+b-1)/[(b-1)^2(b-2)] = \mu\left(\frac{a+1}{b-2} - \mu\right)$	
Mode	$(a-1)/(b+1)$	$a > 1$
$E[X/(1-X)]^k$	$\frac{\Gamma(a+k)\Gamma(b-k)}{\Gamma(a)\Gamma(b)}$	
Skewness	$\gamma_1 = 2(b-a)\sqrt{a+b+1}/[\sqrt{ab}(a+b+2)]$	
Kurtosis	$\beta_2 = 3c(c+1)(a+1)(2b-a)/[ab(c+2)(c+3)]$	$c = a+b$
Mean deviation	$E\lvert X-\mu\rvert = 2\int_0^{a/(b-1)} I_{y/(1+y)}(a,b)\,dy$	
Moments	$\mu_r' = B(a+r, b-r)/B(a,b)$	$a^{(r)}/(b)_{(r)}$
Moments	$\mu_r = (-c)^r \,{}_2F_1\,(a,-r,a+b,1/c)$	$c = a/(a+b)$
ChF	${}_1F_1(a,a+b;it) = \frac{\Gamma(a+b)}{\Gamma(a)} \sum_{j=0}^{\infty} \frac{\Gamma(a+j)(it)^j}{\Gamma(a+b+j)\Gamma(1+j)}$	
Additivity	$\sum_{i=1}^{m} \mathrm{Beta}(a_i, b) = \mathrm{Beta}(\sum_{i=1}^{m} a_i, b)$	Independent
Recurrence	$f(x; a+1, b)/f(x; a, b) = (1 + b/a)(x/(1+x))$ $f(x; a, b+1)/f(x; a, b) = (1 + a/b) * 1/(1-x)$	
U-shaped	$a < 1$ and $b < 1$	
J-shaped	$(a-1) * (b-1) < 0$	
Tail area	$I_x(a, b) = \frac{1}{B(a,b)} \int_0^x t^{a-1}(1-t)^{b-1}\,dt$	$x = y/(1+y)$

Beta distribution of the first kind (also called type-I beta distribution) is obtained by the transformation $X = Y/(1 + Y)$.

Equate the numerator to zero and solve for y to get $y[a + b - a + 1] = (a - 1)$, or $y = (a - 1)/(b + 1)$.

As the Beta-I random variable takes values in $[0, 1]$, any CDF can be substituted for x to get a variety of new distributions (Chattamvelli (2012) [36]). For instance, put $x = \Phi(x)$, the CDF of a normal variate to get the beta-normal distribution with PDF

$$f(x; a, b) = (1/B[a, b])\ \phi(x)\,[\Phi(x)]^{a-1}\,[1 - \Phi(x)]^{b-1}, \tag{4.25}$$

where $B(a, b)$ is the CBF, $\phi(x)$ is the PDF, and $\Phi(x)$ the CDF of normal distribution, so that the range is now extended to $-\infty < x < \infty$.

The CDF of type-II beta can be expressed as $B(z; a, b)/B(a, b) = I_z(a, b)$ where $z = x/(1 + x)$.

4.6   SIZE-BIASED BETA DISTRIBUTIONS

As the mean of Beta-I(a, b) distribution is $a/(a + b)$, a size-biased Beta-I distribution can be obtained as

$$f(x; a, b) = (a + b)x^a (1 - x)^{b-1}/[a \, B(a, b)] = x^a (1 - x)^{b-1}/B(a + 1, b), \qquad (4.26)$$

which belongs to the same family. Alternately, consider the linear function $(1 + cx)$ with expected value $1 + ac/(a + b)$ from which $f(x; a, b, c) = (a + b)(1 + cx)x^{a-1}(1 - x)^{b-1}/[a(a + b + ac) \, B(a, b)]$, which simplifies to $f(x; a, b, c) = (1 + cx)x^{a-1}(1 - x)^{b-1}/[(a + b + ac) \, B(a + 1, b)]$. Similarly, size-biased Beta-II distribution can be obtained as

$$f(x; a, b) = (b - 1)x^a/[a B(a, b)(1 + x)^{a+b}] = x^a/[B(a + 1, b - 1)(1 + x)^{a+b}] \qquad (4.27)$$

which is again a Beta-II distribution.

4.7   THE INCOMPLETE BETA FUNCTION

The incomplete beta function (IBF) denoted by $I_x(a, b)$ or $I(x; a, b)$ has several applications in statistics and engineering. It is used in wind velocity modeling, flood water modeling, and soil erosion modeling. It is used to compute Bartlett's statistic for testing homogeneity of variances when unequal samples are drawn from normal populations [64], and in several tests involving likelihood ratio criterion [114]. It is also used in computing the power function of nested tests in linear models [122], in approximating the distribution of largest roots in multivariate inference. Its applications to traffic accident proneness are discussed by Haight (1956) in [71].

4.7.1   TAIL AREAS USING IBF

Tail areas of several statistical distributions are related to the beta CDF, as discussed below. The survival function of a binomial distribution BINO(n, p) is related to the left tail areas of Beta-I distribution as:

$$\sum_{x=a}^{n} \binom{n}{x} p^x q^{n-x} = I_p(a, n - a + 1). \qquad (4.28)$$

Using the symmetry relationship, the CDF becomes

$$\sum_{x=0}^{a-1} \binom{n}{x} p^x q^{n-x} = I_q(n - a + 1, a). \qquad (4.29)$$

When both a and b are integers, this has a compact representation as

$$I_x(a,b) = 1 - \sum_{k=0}^{a-1} \binom{a+b-1}{k} x^k (1-x)^{a+b-1-k}. \tag{4.30}$$

The survival function of negative binomial distribution is related as follows:

$$\sum_{x=a}^{n} \binom{n+x-1}{x} p^n q^x = I_q(a,n) = 1 - I_p(n,a). \tag{4.31}$$

The relationship between the CDF of central F distribution and the IBF is

$$F_{m,n}(x) = I_y\,(m/2,n/2)\,, \tag{4.32}$$

where (m,n) are the numerator and denominator DoF and $y = mx/(n+mx)$. Similarly, Student's t CDF is evaluated as

$$T_n(t) = (1/2)\left(1 + \text{sign}(t)\, I_x(1/2,n/2)\right) = (1/2)\left(1 + \text{sign}(t)\,[1 - I_y(n/2,1/2)]\right), \tag{4.33}$$

where $x = t^2/(n+t^2)$, $y = 1 - x = n/(n+t^2)$, $\text{sign}(t) = +1$ if t$>$ 0, -1 if t$<$ 0 and is zero for $t = 0$.

The IBF is related to the tail areas of binomial, negative binomial, Student's t, central F distributions. It is also related to the confluent hypergeometric function, generalized logistic distribution, the distribution of order statistics from uniform populations, and the Hotelling's T^2 statistic. The hypergeometric function can be approximated using the IBF also [145]. The Dirichlet (and its inverse) distribution can be expressed in terms of IBF [140]. It is related to the CDF of noncentral distributions. For instance, the CDF of singly noncentral beta (Seber (1963) [121]), singly type-II noncentral beta, and doubly noncentral beta (Chattamvelli (1995) [31]), noncentral T (Chattamvelli (2012) [36], Craig (1941) [48]), noncentral F (Chattamvelli (1996) [33], Patnaik (1949) [107]), and the sample multiple correlation coefficient (Ding and Bargmann (1991) [53], Ding (1996) [52]) could all be evaluated as infinite mixtures of IBF. It is used in string theory to calculate and reproduce the scattering amplitude in terms of Regge trajectories, and to model properties of strong nuclear force.

Definition 4.16 The IBF is the left-tail area of the Beta-I(a,b) distribution

$$I_x(a,b) = (1/B\text{(a,b)}) \int_0^x t^{a-1}(1-t)^{b-1} dt, \quad (a,b > 0) \text{ and } 0 \le x \le 1, \tag{4.34}$$

where $B(a,b)$ is the CBF. Obviously, $I_0(a,b) = 0$, and $I_1(a,b) = 1$. Replace x by $(1-x)$ and swap a and b to get a symmetric relationship

$$I_x(a,b) = 1 - I_{1-x}(b,a). \tag{4.35}$$

This symmetry among the tail areas was extended by Chattamvelli (1995) [31] to noncentral beta, noncentral Fisher's Z, and doubly noncentral distributions. We write $I_x(b)$ or $I(x;b)$ for the symmetric IBF $I_x(b,b)$. The parameters of an IBF can be any positive real number. Simplified expressions exist when either of the parameters is an integer or a half-integer. These representations have broad range of applications to evaluating or approximating other related distributions and test statistics.

The IBF has representations in terms of other special functions and orthogonal polynomials. For example, it could be expressed in terms of hypergeometric series as

$$I_x(a,b) = \frac{x^a(1-x)^{b-1}}{aB(a,b)}\ {}_2F_1(1-b,1;a+1;-x/(1-x)),\qquad(4.36)$$

where ${}_2F_1$ denotes the hypergeometric series.

4.7.2   RANDOM NUMBERS

Random variate generation from beta distribution is accomplished using the relationship between the beta and gamma distributions. Hence, if 2 random numbers are generated from GAMMA(1,a) and GAMMA(1,b) where a<b, then the beta variate is given by B(a,b)=GAMMA(1,a)/[GAMMA(1,a)+GAMMA(1,b)].

4.7.3   FITTING

Estimates of parameters cannot be obtained directly because the mean and variance are nonlinear functions of the parameters. Modified MoM estimation technique can be use to approximate the estimation process. Exact estimates are available only for particular parameter values (Cribari-Neto and Vasconcellos (2002) [49]). MLE estimation involves a pair of digamma functions as $\psi(a)-\psi(a+b) = (1/n)\sum_{k=1}^{n}\log(x_k)$ and $\psi(b)-\psi(a+b) = (1/n)\sum_{k=1}^{n}\log(1-x_k)$. If it can be inferred from a plot of the data that the likely values of a and b are large, an approximation of digamma function as $\psi(x) \approx \log(x-0.5)$ may be used.

4.8   GENERAL BETA DISTRIBUTION

General three-parameter beta distribution is given by

$$f_x(a,b,c) = (x/c)^{a-1}(1-x/c)^{b-1}/c\ B(a,b).\qquad(4.37)$$

The four-parameter beta distribution follows from (4.1) using $y = (x-a)/(b-a)$ as

$$f(x;a,b,c,d) = \frac{\Gamma(c+d)}{\Gamma(c)\Gamma(d)(b-a)^{c+d-1}}(x-a)^{c-1}(b-x)^{d-1}.\qquad(4.38)$$

This could also be written as

$$f(x;a,b,c,d) = \frac{\Gamma(c+d)}{\Gamma(c)\Gamma(d)(b-a)}[(x-a)/(b-a)]^{c-1}[1-(x-a)/(b-a)]^{d-1},\qquad(4.39)$$

which can be transformed to Beta-I using $y = (x-a)/(b-a)$. This has mean $(ad+bc)/(c+d)$, and variance $\sigma^2 = cd(b-a)^2/[(c+d+1)(c+d)^2]$. The location parameters are "a", "b" and scale parameters are c and d. Coefficient of skewness is $2cd(d-c)/[(c+d)^2(c+d)^{(3)}[cd/((c+d)(c+d)^{(2)})]]$ where $(c+d)^{(k)}$ is raising Pochhammer notation with $(c+d)^{(3)} = (c+d)(c+d+1)(c+d+2)$. The mode is $\frac{a(d-1)+b(c-1)}{(c+d-2)}$ for c not 1 and d not 1.

The beta-geometric (discrete) distribution is defined in terms of CBF as

$$f(x;a,b) = B(a+1, x+b-1)/B(a,b), \quad \text{for x} = 1,2,3,\ldots \quad (4.40)$$

This satisfies the recurrence relation $(a+b+x-1)p_x(a,b) = (x+b-2)p_{x-1}(a,b)$ with $p_0 = B(a+1,b-1)/B(a,b)$. A change of origin transformation $Y = X - 1$ results in the PMF $f(x;a,b) = B(a+1, x+b)/B(a,b)$, for x = 0,1,2,\ldots

4.8.1 ADDITIVITY PROPERTY

If X_k are IID Beta-I(a_k, b) random variables, then $Y = \sum_k X_k$ have the same distribution with parameters $(\sum_k a_k, b)$ (i.e., $\sum_{i=1}^m$ Beta-I(a_i,b) = Beta-I$(\sum_{i=1}^m a_i, b)$).

4.9 APPLICATIONS

Approximate beta distribution is assumed in several engineering and applied science fields. For example, shear strength parameters in geotechnical engineering (cohesive force c and internal friction angle f), path-averaging time in boundary layer meteorology (scintillometer measurements of the inner scale (x) and the refractive index structure parameter (y)), uncertain solar irradiance in photovoltaic power generation [13], distribution of orbital eccentricities for exoplanets()[3] are all assumed to follow an approximate beta distribution. A beta distribution of the mixture fraction is assumed in single-feed semi-batch stirred-tank reactors used in crystallography, pharmaceuticals and other manufacturing processes that involve chemical reactions and precipitation processes [23]. Modeling the loss given default of portfolio credit risk uses Beta-I distribution. Some geometric applications of the beta distribution can be found in [62], and an application for drinking water data analysis in [132].

4.9.1 CIVIL ENGINEERING

It is used as an approximation to the distribution of proportions (like fraction of a highway stretch that requires repair works, proportion of total work involved in a construction project, etc.), and to model geotechnical parameters. It is also used in construction simulation modeling, reliability centered maintenance, and criticality analysis (Nadarajah and Kotz (2007) [102]). Trustworthy forecasting is instrumental in successful completion of large and complex projects. Resource wastage (materials, man-power, machines, etc.) can be kept to a minimum when pessimistic, average, and optimistic estimates on project completion are available. The principle is to

---

[3]Another choice to model exoplanet eccentricity is a mixture of Rayleigh and exponential distribution.

predict the minimum, average (usually using the mode), and maximum activity durations using deterministic methods.

4.9.2   GEOTECHNICAL ENGINEERING

The shear strength parameters in geotechnical engineering (cohesive force c, and internal friction angle ϕ) are crucial in accurate reliability analysis. The risk assessment accuracy can then be modeled using a joint distribution of c and ϕ. Data scarcity may lead to inaccurate estimates of the probability of failure. Either a truncated normal, half-normal, truncated lognormal[4] or a beta distribution (with range $[a, b]$) is assumed for the above parameters. As there are multiple parameters (like cohesive force, internal friction angle, unit weight of soils) involved, one approach is to approximate the joint distribution by a univariate distribution. This is called the "copula-approach," or "copula modeling technique." As the shear strength parameter is more important to achieve high accuracy, marginal distribution of it using the beta law can improve the accuracy of reliability analysis. Restricting attention to only the shear-strength (c) and internal friction angle(ϕ), the bivariate CDF $F(c, \phi)$ can be expressed in terms of individual marginal distributions and a copula function as

$$F(c, \phi) = C(F_1(c), F_2(\phi); \theta), \qquad (4.41)$$

where $C()$ denotes the copula. Take partial derivative $\partial^2/\partial c\partial\phi$ to get

$$f(c, \phi) = c(F_1(c), F_2(\phi); \theta) f_1(c) f_2(\phi), \qquad (4.42)$$

where $f_1(c)$ and $f_2(\phi)$ are the marginal PDFs. Some geotechnical processes are multi-modal (exhibit two or more distinct peaks) in which case linear combination of appropriate uni-modal distributions are used in reliability analysis under uncertainties.

4.9.3   BETA DISTRIBUTION IN CONTROL CHARTS

Quantitative reliability modeling leads to more effective prediction of product failures and could improve work planning. Consider an acceptance sampling plan with nonconforming fraction p. If the process mean and standard deviation are unknown, they are estimated using minimum variance unbiased (MVU) technique. Acceptance sampling by variables (ASV) is a technique used by modern SQC systems to identify process deviations. Univariate ASV provide acceptance tests based on MVU estimate of the process fraction nonconforming, p.

4.9.4   BETA DISTRIBUTION IN PERT

The program (or project) evaluation and review technique (PERT) is a diagrammatic tool used in project management. It was first introduced in 1957 for the U.S. Navy's Polariz nuclear submarine design and construction scheduling project. The project must be comprised of tasks (called

---

[4]As the soil properties are strictly non-negative, the lognormal is preferred over normal distribution.

activities) with a dependency among them. Each activity is uniquely identified using a start and end dates (or times in micro-projects), and represented by an arrow. Isolated activities that do not have dependency among other activities are excluded from PERT. This implies that the PERT graph is always a directed acyclic graph (DAG) with the project start-date as the start-node (or source), and project finish-date as the end-node (sink) with predecessor and successor events for all intermediate activities.[5] Its primary purpose is to analyze various activities so as to provide a best and worst estimates on project completion time and costs. In other words, uncertainty is incorporated in a controlled manner so that projects can be scheduled without knowing the precise details and durations of all the activities involved. The information on early-start (ES), early-finish (EF), late-start (LS), late-finish (LF), and expected duration can be obtained for internal nodes (and sink node) so that management can schedule activities in an optimal way (manpower, materials, machines, etc.) to complete a project within constraints. A critical path (which is the path with the longest time to complete) is identified from the source to the sink which identifies all activities with slack. Even internal nodes can be analyzed to understand each completed phase of a complex project, so that management can periodically review the progress within scheduled time and cost expenditures. A similar tool called critical path method (CPM) is also popular in project management. Although PERT and CPM are complementary tools, CPM uses *one time* and *one cost* estimation for each activity, so that PERT is more versatile for analysis of milestones in big projects.

PERT uses four types of time estimates to accomplish an activity. An optimistic-estimate (o) is the minimum possible time required, a pessimistic-estimate (p) is the maximum possible time required, a most-likely time (m) is the best estimate of the time required (mode), and an expected time (o+4m+p)/6 is the average (arithmetic mean) time required, with variance $(p - o)^2/36$. Activity duration in PERT networks (used in project planning and implementations) are assumed to follow the beta distribution, in which case more precise estimates are available for expected time as (2o+9m+2p)/13. It may also be associated with any particular set of PERT estimates. The four-parameter beta distribution is typically used in PERT modeling (especially to model earth-moving activities in construction projects). The PDF is given by

$$f(x; a, b, p, q) = (x - c)^{a-1}(d - x)^{b-1}/[(d - c)^{a+b-1}B(a, b)], \qquad (4.43)$$

where c (most optimistic completion time) is the lower and d (most pessimistic completion time) is the upper limit on activity duration.

4.9.5   CONFIDENCE INTERVALS FOR BINOMIAL PROPORTIONS

Its applications in statistics include finding distribution-free tolerance intervals, and confidence intervals for proportions. Let p_l and p_u denote the lower and upper confidence intervals for the proportion p of a binomial distribution. The exact $100(1-\alpha)\%$ CI for a binomial propor-

---

[5]Nodes represent activities and arcs represent precedence relationship between them.

tion is $100\alpha/2$-th percentile of beta distribution Beta-I$(x, n - x + 1)$ for p_L, and $100(1-\alpha/2)$-th percentile of Beta-I$(x + 1, n - x)$ for p_U. It can be shown using the relation between the sum of tail probabilities of binomial and tail areas of beta distributions that acceptance tests in several manufacturing environments are built on the MVU estimate of p, the process fraction nonconforming. The MVU estimate of p beyond a given specification limit is given by $I_c(n/2 - 1, n/2 - 1)$ where $c = \frac{1}{2}(1 - (U - \overline{x})\sqrt{n}/((n - 1)s))$ for an upper specification limit U, and $c = \frac{1}{2}(1 - (\overline{x} - L)\sqrt{n}/((n - 1)s))$ for a lower specification limit L.

4.9.6   MINERALOGY

Many methods exist for mineral extraction from ores or raw materials. One popular method called ore-dressing uses physical, chemical, magnetic, other properties of materials to separate the minerals. This process of separation at particulate level is called "liberation" in mineralogy. The four- and six-parameter versions of beta distribution are used in mineralogy and mining engineering for this purpose.

4.10   SUMMARY

This chapter introduced different types of beta distributions. Despite its versatility in modeling a variety of practical problems, it is not easy to fit because the mean and variance are nonlinear functions of the parameters. The IBF that has wide-ranging applications in various fields is also discussed.

CHAPTER 5

Arcsine Distribution

5.1   INTRODUCTION

The arcsine distribution has lots of applications in diverse areas like Brownian motion, actuarial sciences, fiducial inference, financial services, statistical linguistics, stochastic processes, thermal calibration systems, etc. It is used to model rates and proportions of various types. This distribution gets its name from the fact that the CDF is $F(x) = (2/\pi)\arcsin(\sqrt{x})$ (inverse of sin() or hyperbolic sine function). It is a continuous distribution that belongs to the trigonometric family. The *standard* arcsine distribution (SASD) of first kind has support $0 < x < 1$, is U-shaped, and symmetric around $x = 1/2$ (see Figure 5.1). Its PDF is given by

$$f(x) = \begin{cases} \frac{1}{\pi\sqrt{x(1-x)}} & \text{for} \quad 0 < x < 1 \\ 0 & \text{elsewhere.} \end{cases}$$

To prove that this is indeed a PDF, put $x = \sin^2(\theta)$ so that $dx = 2\sin(\theta)\cos(\theta)d\theta$, and $1 - x = 1 - \sin^2(\theta) = \cos^2(\theta)$. The denominator $\sqrt{x(1-x)}$ becomes $\sqrt{\sin^2(\theta)\cos^2(\theta)} = \sin(\theta)\cos(\theta)$. When $x = 0$, $\theta = 0$, and when $x = 1$, $\theta = \pi/2$.

$$\int_0^{+1} \frac{1}{\pi\sqrt{x(1-x)}} dx = \int_0^{\pi/2} (2/\pi)\sin(\theta)\cos(\theta)d\theta/[\sin(\theta)\cos(\theta)]$$

$$= (2/\pi)\int_0^{\pi/2} d\theta = (2/\pi)\theta|_0^{\pi/2} = 1. \tag{5.1}$$

This shows that it is indeed a PDF. The above distribution is denoted as SASD-I. Another form of the distribution called "standard arcsine distribution of second kind" (SASD-II) or "centered arcsine distribution" (CASD) has support $-1 < x < 1$ with PDF given by

$$f(x) = \begin{cases} \frac{1}{\pi\sqrt{(1-x^2)}} & \text{for} \quad -1 < x < 1 \\ 0 & \text{elsewhere.} \end{cases}$$

It is called inverse sine distribution, Chebyshev distribution on $(-1, 1)$ in approximation theory, Haldane prior in astronomy, and Kesten–McKay distribution (on $(-2, 2)$) in spectral graph theory.[1]

---

[1]Note that these are all open intervals because the density $\to \infty$ at the extreme points. That is why we used simple parentheses instead of square brackets $[-1, +1]$, etc. Most of the authors in other fields seem to ignore this fact.

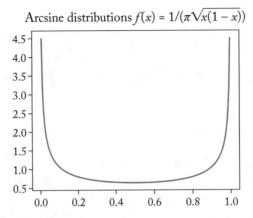

Figure 5.1: Standard arcsine distribution.

To prove that this is a PDF, integrate over the range to get $\int_{-1}^{+1} 1/\sqrt{(1-x^2)}dx = \sin^{-1} x|_{-1}^{+1} = (3\pi/2) - (\pi/2) = \pi$. The π cancels out, showing that this is indeed a PDF. A change of scale transformation results in the distribution

$$f(x; b) = 1/[2\pi \sqrt{(b^2 - x^2)}], \quad \text{for} \quad 0 < x < b, \tag{5.2}$$

which is used in "arcsine transforms" of monotonic functions (page 66).

Problem 5.1   If $X \sim$ SASD-II with PDF $f(x) = 1/[\pi \sqrt{(1-x^2)}]$, find the expected value of $\log(X - Y)^2$ where $Y \in [-1, +1]$.

5.1.1   ALTERNATE REPRESENTATIONS

The standard arcsine distribution SASD-I has range $(0, 1)$, and SASD-II has range $(-1, +1)$. They are also written as $\frac{1}{\pi}(x(1-x))^{-1/2}$ (or $(\pi^2 x(1-x))^{-1/2}$, $(\pi^2 x(1-x))^{-0.5}$) and $\frac{1}{\pi}(1 - x^2)^{-1/2}$ (or $(\pi^2(1-x^2))^{-1/2}$, $(\pi^2(1-x^2))^{-0.5}$), respectively. As $\Gamma(1/2) = \sqrt{\pi}$, it can also be written as $(x(1-x))^{-1/2}/\Gamma(1/2)^2$. Arcsine distribution with PDF

$$f(x; R) = 1/[\pi \sqrt{x(R-x)}], \quad \text{for} \quad 0 < x < R, \tag{5.3}$$

is used in Weiner processes [120]. The type-II analogue is

$$f(x; R) = R/[\pi \sqrt{(R^2 - x^2)}], \quad \text{for} \quad -R < x < R. \tag{5.4}$$

A question that naturally arises is whether $f(x; R) = R/[\pi \sqrt{(R^2 + x^2)}]$ is a valid PDF. In fact, this is also a PDF for $x > 0$ whose CDF is the inverse hyperbolic function $(1/R) \sinh^{-1}(x/R)$. Geometrically, the density on $[0, 1]$ (SASD-I) gives the distribution of the projection of a random point on a circle of radius half centered at $(0.50, 0)$ to the continuous interval $[0, 1]$ on the X-axis, and as projection of a random point on a centered circle (at origin) with appropriate radius for symmetric versions (e.g., SASD-II). When the domain is $[-R, +R]$ this circle is origin-centered with radius R. Shifts of the circle on the horizontal axis results in other displaced distributions discussed below. By assuming that the circle rolls continuously at constant speed horizontally, it can be used to model the position of a particle moving in simple harmonic motion with amplitude R at a random time t. It is also used in von Neumann algebra theory. The two-parameter ASD-I has PDF

$$f(x; a, b) = 1/[b\pi \sqrt{((x-a)/b)(1-(x-a)/b)}] \quad \text{for} \quad a < x < a+b, \qquad (5.5)$$

and the corresponding ASD-II has PDF

$$f(x; a, b) = 1/[b\pi \sqrt{1 - ((x-a)/b)^2}] \quad \text{for} \quad a - b < x < a+b. \qquad (5.6)$$

This is a location-and-scale distribution that is symmetric around a and is U-shaped. It reduces to the SASD by the transformation $Y = (X-a)/b$. Differentiate w.r.t. x, and equate to zero to get the minimum at $x = a$, with minimum value $f(a) = 1/(b\pi)$. In terms of the minimum value, the PDF (5.6) can be written as

$$f(x; a, b) = f(a)/\sqrt{1 - ((x-a)/b)^2} \quad \text{for} \quad a - b < x < a+b. \qquad (5.7)$$

Next, consider the PDF

$$f(x; a, b) = 1/[\pi \sqrt{(x-a)(b-x)}] = [\pi^2 (x-a)(b-x)]^{-1/2} \quad \text{for} \quad a < x < b, \qquad (5.8)$$

which has applications in sinusoidal data modeling. It reduces to SASD-I for $a = 0$ and $b = 1$ (or $b = 0$ and $a = 1$), and to SASD-II for $a = -1$ and $b = +1$ (or $b = -1$ and $a = 1$). There are many ways to prove that this is indeed a PDF. We could write the square root expression as $-(x-a)(x-b) = -(x^2 - (a+b)x + ab)$, complete the square to get this in the form $(a^2 + b^2)/4 - ab/2 - (x - (a+b)/2)^2 = [(a-b)/2]^2 - (x - (a+b)/2)^2$, and then use the standard integration rule $\int_a^b dy/\sqrt{c^2 - y^2} = (1/c) \sin^{-1}(y/c)|_a^b$ where $c = (a-b)/2$, and ranges are modified accordingly. There is an alternate simple way to integrate the above. Put $y = x - a$, so that the limits are from 0 to $b - a$. The integrand becomes $1/[\pi \sqrt{y(c-y)}]$ where $c = b - a$. Now put $y = c \sin^2(\theta)$ so that $dy = 2c \sin(\theta) \cos(\theta)$. When $y = b - a$, $\theta = \sin^{-1}(1) = \pi/2$ (because $c = b - a$ cancels out). The constant c, and $\sin(\theta) \cos(\theta)$ cancel out from the numerator and denominator. The integral becomes $(2/\pi) \int_0^{\pi/2} d\theta = 1$. Equation (5.8) is written in some fields as

$$f(x; a, b) = 1/[\pi \sqrt{(x-a)(a+b-x)}] = [\pi^2 (x-a)(a+b-x)]^{-0.5} \quad \text{for} \quad a < x < a+b. \qquad (5.9)$$

5.1.2   RELATION TO OTHER DISTRIBUTIONS

This is a special case of the beta distribution (Chapter 4) when $a = 1/2, b = 1/2$. Hence, all properties of beta distribution are applicable to SASD-I as well. In particular, X and $1 - X$ are identically distributed. As the range of SASD-I is (0,1), the transformation $Y = -\log(X)$ results in log-arcsine distribution discussed in page 60. If X'_ks are IID Beta-I$(\frac{2k-1}{2n}, \frac{1}{2n})$ random variables, the distribution of the geometric mean (GM) of them $Y = (\prod_{k=1}^{n} X_k)^{1/n}$ is SASD-I distributed for $n \geq 2$ ([90], [22]). This has the interpretation that the log-arcsine law is decomposable into a sum (or average) of independent log-beta random variables (so that arcsine distribution is not additively decomposable or infinite divisible). If U is CUNI(0, 1) then $Y = -\cos(\pi U/2)$ is arcsine distributed. Conversely, if X has an arcsine distribution, $U = (2/\pi) \arcsin(\sqrt{x})$ has the $U(0, 1)$ distribution. Differentiate w.r.t. u to get $|\partial y/\partial u| = (\pi/2) \sin(\pi u/2)$, so that $|\partial u/\partial y| = (2/\pi) / \sqrt{1 - \cos^2(\pi u/2)} = (2/\pi)/\sqrt{1 - y^2}$. An alternate way to state this is as follows. If $U \sim$ CUNI$(-\pi, \pi)$, the distribution of $Y = \cos(u)$ is SASD-I (see below). Similarly, the SASD-II is related to the $U(0, 1)$ distribution as $X = \cos(\pi u)$, because $|\partial x/\partial u| = \pi \sin(\pi u) = \pi \sqrt{1 - \cos^2(\pi u)} = \pi \sqrt{1 - x^2}$. The transformation $Y = 2X - 1$ and $Y = X^2$ when applied to SASD-II results in SASD-I. Similarly, if $X \sim$ SASD-I, then $Y = \sqrt{X}$ is SASD-II. If $\Phi()$ denotes the CDF of a normal distribution, $\Phi^{-1}(F(x)) \sim$ N(0,1) where $F(x)$ denotes the CDF of ASD.

Problem 5.2   If $X \sim$ SASD-I prove that $Y = 1/X$ has PDF $f(y) = 1/[\pi y \sqrt{y - 1}]$ for $y > 1$.

Example 5.3   Distribution of cos(X)   If $X \sim$ CUNI$(-\pi, \pi)$, find the distribution of $Y = \cos(X)$.

Solution 5.4   As $X \sim$ CUNI$(-\pi, \pi)$, $F(x) = 1/2\pi$. From $y = \cos(x)$, we get $|dy/dx| = \sin(x) = \sqrt{1 - \cos^2(x)} = \sqrt{1 - y^2}$, so that $f(y) = (1/2\pi)(1/\sqrt{1 - y^2})$. Since the equation $y = \cos(x)$ has two solutions in $-\pi, \pi$ as $x_1 = \cos^{-1}(y)$ and $x_2 = 2\pi - x_1$, the PDF becomes $f(y) = 1/(\pi \sqrt{1 - y^2})$, which is SASD-II.

Example 5.5   Transformation of Arcsine distribution   If X is Arcsine distributed, find the distribution of $Y = -\log(X)$.

Solution 5.6   Let $F(y)$ be the CDF of Y. Then $F(y) = \Pr[Y \leq y] = \Pr[X \geq \exp(-y)]$ $= 1 - (2/\pi) \sin^{-1}(\sqrt{\exp(-y)})$. Differentiate w.r.t. y to get the PDF as $f(y) = (2/\pi) \exp(-y/2)/[2(\sqrt{1 - \exp(-y)})]$. The 2 cancels out from numerator and denominator giving

$$f(y) = (1/\pi) \exp(-y/2)/\sqrt{1 - \exp(-y)}, \quad 0 \leq y < \infty. \tag{5.10}$$

5.2   PROPERTIES OF ARCSINE DISTRIBUTION

The SASD-I is a special case of beta type-I distribution. It is symmetric around the mean (1/2) and is concave upward (the probability decreases and then increases), but satisfies the log-convex property. As $x \to 0$ or $x \to 1$ the PDF $f(x) \to \infty$. Put $Y = X - \frac{1}{2}$ to get

$$f(y) = \frac{1}{\pi \sqrt{(y + 1/2)(1/2 - y)}}, \quad -1/2 < y < 1/2. \tag{5.11}$$

As $(1/2 + y)(1/2 - y) = (1/4 - y^2)$, the PDF becomes $f(y) = (2/\pi)/\sqrt{(1 - 4y^2)}$, for $-1/2 < y < 1/2$. The mean is 0.5 and variance is 0.125 for the standard arcsine distribution (see below). As the distribution is symmetric, coefficient of skewness is zero. The kurtosis coefficient is $\beta_2 = 3/2$. Thus, it is always platykurtic. Note that the density is maximum when x is near 0 or 1 with the center as a cusp (U-shaped), and minimum at $x = 0.5$ with minimum value $2/\pi$. Hence, there are two modes (bimodal) that are symmetrically placed in the tails. This is the reason why it is platykurtic.

Arcsine distribution is the exact antithesis of bell-shaped laws because (i) the mean coincides with the minimum (whereas mean coincides with maximum for bell-shaped laws), (ii) lower and upper limits correspond to asymptotes (density rises up to ∞) (whereas bell-shaped laws tail off to zero), (iii) bimodal (bell-shaped laws are unimodal), and (iv) statistical measures are more prone to outliers as the peaks are away from the mean (samples from bell-shaped distributions have lesser chance of outliers). Due to these peculiarities, the convergence of central limit theorem to normality is slow. The hazard function is given by

$$1/h(x; a, b) = b \sqrt{1 - [(x - a)/b]^2} \; \arccos((x - a)/b). \tag{5.12}$$

The two-parameter ASD satisfies an interesting property:– If $X \sim \text{ASD}(a, b)$ then $cX + d \sim \text{ASD}(ac + d, bc + d)$. See Table 5.1 for further properties.

5.2.1   MOMENTS OF ARCSINE DISTRIBUTION

As the distribution is symmetric, all odd central moments higher than 1 vanish. Ordinary moments (around zero) of this distribution are easy to find.

Example 5.7   Mean of arcsine distributions   Find the mean of arcsine distribution of the first kind.

Solution 5.8 $E(X) = (1/\pi) \int_0^1 x/\sqrt{x(1 - x)}dx$. Put $x = \sin^2(\theta)$ as before so that $E(X) = (2/\pi) \int_0^{\pi/2} \sin^2(\theta)d\,\theta$. Put $\sin^2(\theta) = (1 - \cos(2\theta))/2$ and integrate to get $2/\pi[(1/2)\theta|_0^{\pi/2} - (1/4)\sin(2\theta)|_0^{\pi/2}] = 2/\pi[\pi/4 - 0] = 1/2$.

Table 5.1: Properties of arcsine distribution

Property	Expression	Comments
Range of X	$0 < x < 1$	SASD-I; continuous
Mean	$\mu = 1/2 = 0.50$	
Median	0.50	Mode $\in \{0, 1\}$
Variance	$\sigma^2 = 1/8$	0.125
Skewness	$\gamma_1 = 0$	Symmetric
Kurtosis	$\beta_2 = 3/2$	Always platykurtic
Mean deviation	$E\lvert X - \mu\rvert = 1/\pi$	0.31831
CV	$1/\sqrt{2}$	
CDF	$\frac{2}{\pi}\sin^{-1}(\sqrt{x})$	$\sin^{-1}(x) + \pi/2$
Moments	$(1/\pi)\mathrm{B}(k + 0.5, 0.5)$	$\mu_{2k} = \binom{2k}{k}(1/2)^{2k}$
MGF	$e^{t/2}\mathrm{I}_0(t/2)$	Modified Bessel function
ChF	$e^{-it/2}\mathrm{I}_0(it/2)$	$_1F_1(1/2, 1; it)$

The mean, median, and skewness coefficient of SASD-II $f(x) = \dfrac{1}{\pi\sqrt{(1-x^2)}}$ are zeros. The variance is $1/2$ and kurtosis coefficient is $-3/2$. Raw moments of SASD-I is

$$\mu_n' = E(X^n) = \prod_{j=0}^{n-1}(2j + 1)/(2j + 2) = (1/2^n)\prod_{j=0}^{n-1}(2j + 1)/(j + 1). \qquad (5.13)$$

Even central moments of arcsine distribution of second kind is $\mu_{2k} = \binom{2k}{k}(1/2)^{2k}$ (odd central moments are zeros due to symmetry). The MGF is

$$M_x(t) = e^{t/2}\mathrm{I}_0(t/2), \qquad (5.14)$$

where $\mathrm{I}_0(x)$ is the modified Bessel function of first kind. The MGF of SASD-II is

$$M_x(t) = \sum_{k=0}^{\infty} t^{2k}(2k - 1)!!/(2k!!), \qquad (5.15)$$

where $n!! = n(n - 2)(n - 4)\ldots(2 \text{ or } 1)$.

Example 5.9 k^{th} moment of arcsine distribution   Find the kth moment of arcsine distribution of second kind.

Solution 5.10   The kth moment is

$$\mu_k' = (\frac{1}{\pi}) \int_{-1}^{+1} x^k / \sqrt{(1-x^2)} dx. \tag{5.16}$$

Put $x = \sin(\theta)$ so that $dx = \cos(\theta)d\theta$, and $\sqrt{(1-x^2)} = \sqrt{1-\sin^2(\theta)} = \cos(\theta)$. Thus, $\mu_k' = \frac{1}{\pi} \int_{-\pi/2}^{\pi/2} \sin^k \theta d\theta$. As $\sin(\theta)$ is an odd function, the integral vanishes when k is odd. When k is even, the integral becomes $\mu_k' = (\frac{2}{\pi}) \int_0^{\pi/2} \sin^k \theta d\theta$. Using integration by parts this reduces to $(\frac{2}{\pi})(\sqrt{\pi}/2)\Gamma((k+1)/2)/\Gamma(k/2+1) = \Gamma((k+1)/2)/(\sqrt{\pi}\Gamma(k/2+1))$. For $k = 2$ this becomes $(1/\sqrt{\pi})\Gamma(3/2) = 1/2$ by using $\Gamma(1/2) = \sqrt{\pi}$.

This distribution has an interesting property that the mean (and median) coincides with the minimum of the density function. An interpretation of this in Brownian motion is that trajectories with long winning or losing streak is more likely than values near the average [136].

Another two-parameter version is obtained by putting $y = (x - a)/b$ in (5.1) as

$$f(x; a, b) = \begin{cases} \frac{b}{\pi\sqrt{(x-a)(a+b-x)}} & \text{for} \quad a < x < b \\ 0 & \text{elsewhere.} \end{cases}$$

It has mean = median = $(a + b)/2$, variance $(b - a)^2/8$, skewness = 0 and excess kurtosis $-3/2$.

Problem 5.11   Find the minima of the PDF $f(x; a, b) = 1/[\pi \sqrt{(x-a)(b-x)}] = [\pi^2(x-a)(b-x)]^{-1/2}$ and the minimum value attained.

5.2.2   TAIL AREAS

The CDF of SASD-I is

$$F(x) = \int_0^x \frac{1}{\pi\sqrt{x(1-x)}} dx = \frac{2}{\pi} \sin^{-1}(\sqrt{x}) = \frac{2}{\pi} \sin^{-1}(x^{1/2}), \quad 0 < x < 1. \tag{5.17}$$

As the SASD-I is symmetric around $x = 1/2$, this could also be written as

$$F(x) = \frac{1}{2} + \sin^{-1}(2x - 1)/\pi, \quad 0 < x < 1. \tag{5.18}$$

This can be represented in terms of hypergeometric functions as $F(x) = \frac{2}{\pi} \sqrt{x}_2 F_1(1/2, 1/2, 3/2; x) = \frac{2}{\pi} \sqrt{x(1-x)}_2 F_1(1, 1, 3/2; x)$. From this, the quantile function of SASD-I follows as $G^{-1}(p) = \sin^2(\pi p/2)$ (or $G^{-1}(p) = (\sin(\pi(p - .5)) + 1)/2)$ for $p \in [0, 1]$. Particular values are $Q_1 = \sin^2(\pi/8) = (2 - \sqrt{2})/4 \approx 0.1464$, and $Q_3 = \sin^2(3\pi/8) = (2 + \sqrt{2})/4 \approx 0.8536$. As the SASD-II is symmetric around $x = 0$, its CDF can be written as

$$F(x) = \frac{2}{\pi} \sin^{-1}(\sqrt{(x+1)/2}) = 1 - \cos^{-1}(x)/\pi, \quad -1 < x < 1, \tag{5.19}$$

from which its quantile function is $G^{-1}(p) = \cos(\pi(1-p))$. The CDF of $f(x; a, b) = 1/[\pi\sqrt{(x-a)(b-x)}]$ for $a < x < b$ is given by $F(x; a, b) = (2/\pi)\arcsin(\sqrt{(x-a)/(b-a)})$, which could also be written as $(1/\pi)\arccos((a+b-2x)/(b-a))$. Similarly, the CDF of (5.5) is $F(x; a, b) = (2/\pi)\arcsin(\sqrt{(x-a)/b})$ for $x \in [a, a+b]$, from which the quantile function follows as $F^{-1}(p) = a + b\sin^2(\pi p/2)$. The particular values are $Q_1 = a + (b/4)(2 - \sqrt{2})$, $Q_2 = a + b/2$, and $Q_3 = a + (b/4)(2 + \sqrt{2})$.

Problem 5.12 Find the constant C for which $f(x; C) = C/\sqrt{1 - 4x^2}$ is a distribution where $-1/2 < x < 1/2$.

Example 5.13 Mean deviation of arcsine distribution Find the mean deviation of the arcsine distribution using Theorem 1.12 in page 10.

Solution 5.14 We have seen above that the CDF is $\frac{2}{\pi}\sin^{-1}(\sqrt{x})$. As the mean is .5 we get the MD using Theorem 1.12 as

$$\text{MD} = 2\int_{ll}^{\mu} F(x)dx = \frac{4}{\pi}\int_0^{.5} \sin^{-1}(\sqrt{x})dx. \tag{5.20}$$

Put $\sqrt{x} = t$ so that $dx = 2t\,dt$. Adjust the upper limit of integration as $c = \sqrt{.5}$. This gives $\text{MD} = \frac{8}{\pi}\int_0^c t\,\sin^{-1}(t)dt$. Now use $\int t\,\sin^{-1}(t)dt = (2t^2 - 1)/4\sin^{-1}(t) + t\sqrt{1-t^2}/4$. The first term evaluates to zero, and we get the MD as $1/\pi$.

Problem 5.15 The heat flux distribution over a cycle is given by $f(x) = C/\sqrt{1 - (x/a)^2}$, for $-a < x < a$. Find the unknown and the mean. What is the maximum heat flux observable?

5.3 GENERALIZED ARCSINE DISTRIBUTIONS

Several generalized versions of this distribution exist. Most of them are multi-parameter versions. The McDonald arcsine distribution [46] is a three-parameter version with PDF

$$S_x(k; a, b) = 0.5 - \arcsin((x-a)/b)/\pi. \tag{5.21}$$

A three-parameter general arcsine distribution has PDF

$$f(x; a, b, c) = ((x-a)(b-x))^{c-1}/[(b-a)^{2c-1}B(c, c)] \quad \text{for} \quad a < x < b, \tag{5.22}$$

which becomes SASD-II for $c = 1/2$. A shape-generalized SASD-II has PDF

$$f(x; c) = (\sin c\pi)/\pi x^{-c}(1-x)^{c-1} \quad \text{for} \quad 0 < x < 1. \tag{5.23}$$

See [46] for the McDonald arcsine distribution, [47] for another extension of arcsine law. Skewed arcsine distribution has PDF $f(x) = 2g(x)G(ax)$, where $g()$ and $G()$ are the PDF and CDF of arcsine distribution. Size-biased arcsine distributions discussed below are other generalizations.

5.3.1   TRANSMUTED ARCSINE DISTRIBUTIONS

Transmuted distributions were discussed in Chapter 1. The CDF of a transmuted arcsine distribution is given by $G(x;\lambda) = (1+\lambda)F(x) - \lambda F^2(x)$ from which the PDF follows by differentiation as

$$g(x;\lambda) = f(x)[(1+\lambda) - 2\lambda F(x)] = \frac{1}{\pi\sqrt{x(1-x)}}[(1+\lambda) - 4(\lambda/\pi)\arcsin(\sqrt{x})]. \quad (5.24)$$

5.3.2   WEIGHTED ARCSINE DISTRIBUTIONS

The size-biased SASD-I is obtained from the density $f(x;b) = (1+bx)/(\pi\sqrt{x(1-x)})$. Integrate over the range $(0,1)$ to get

$$\int_0^1 f(x;b)dx = \int_0^1 \frac{1+bx}{\pi\sqrt{x(1-x)}}dx, \quad (5.25)$$

where b is a real number. Break this into two integrals and integrate each one to get

$$\int_0^1 f(x;b)dx = 1 + (b/\pi)\int_0^1 x^{1/2}/\sqrt{(1-x)}dx. \quad (5.26)$$

Put $x = \sin^2(\theta)$ so that $dx = 2\sin(\theta)\cos(\theta)d\theta$. The limits for θ are from 0 to $\pi/2$. As the denominator is $\cos(\theta)$, cancel it from numerator. The integral becomes $1 + (b/\pi)\int_0^{\pi/2}2\sin^2(\theta)d\theta$. Write $\sin^2(\theta) = (1-\cos(2\theta))/2$ and integrate to get $1 + (b/\pi)[\pi/2]=1 + b/2$. Thus,

$$f(x;b) = \frac{1+bx}{1+b/2}\frac{1}{\pi\sqrt{x(1-x)}} \quad (5.27)$$

is the size-biased SASD-I distribution.

The size-biased SASD-II is obtained from the density $f(x;b) = (1+bx)/(\pi\sqrt{(1-x^2)})$. Integrate over the range $(-1,1)$ to get

$$\int_{-1}^1 f(x;b)dx = \int_{-1}^1 \frac{1+bx}{\pi\sqrt{(1-x^2)}}dx, \quad (5.28)$$

where b is a real number. Break this into two integrals and integrate each one to get

$$\int_{-1}^1 f(x;b)dx = 1 + (b/\pi)\int_{-1}^1 x/\sqrt{(1-x^2)}dx. \quad (5.29)$$

As this integrand is an odd function, the integral vanishes. Thus, $f(x;b) = (1+bx)/(\pi\sqrt{(1-x^2)})$ is the size-biased SASD-II distribution. Size-biased and weighted versions of other arcsine distributions mentioned above can similarly be obtained. Put $k = 2$ in $\mu'_k = \Gamma((k+1)/2)/(\sqrt{\pi}\Gamma(k/2+1))$ (5.16) to get μ'_2 of SASD-II as $(1/\sqrt{\pi})\Gamma(3/2)/\Gamma(2) = 1/2$. A weighted distribution can be obtained from this as $g(x) = 2x^2 \dfrac{1}{\pi\sqrt{(1-x^2)}}$ for $|x| < 1$. Higher-order weights and convolutions can result in a variety of new distributions.

Example 5.16   If the parameter p of a negative binomial distribution has an arcsine distribution, prove that the resulting mixture distribution is $(1/\pi)\binom{x+k-1}{x}B(k+1/2, x+1/2)$.

Solution 5.17   Consider the PMF

$$f(x; k, p) = \binom{x+k-1}{x}p^k q^x, \tag{5.30}$$

where p is distributed as $f(p) = (1/\pi)/\sqrt{p(1-p)}$. The unconditional distribution is obtained by integrating out p as

$$f(x; k, a, b) = \int_{p=0}^{1} \binom{x+k-1}{x}p^k q^x (1/\pi)/\sqrt{p(1-p)}dp. \tag{5.31}$$

Take constants outside the integral to get

$$f(x; k, a, b) = [1/\pi]\binom{x+k-1}{x}\int_{p=0}^{1} p^{k-1/2}(1-p)^{x-1/2}dp. \tag{5.32}$$

Using complete beta integral this becomes $[B(k+1/2, x+1/2)\binom{x+k-1}{x}]/\pi$.

5.4   APPLICATIONS

The arcsine distribution has applications in random walks, Brownian motion, geostatistics, statistical linguistics, finance, number theory, to name a few. The time at which the minimum of Brownian motion on the interval $[0, 1]$ is achieved, as well as the fraction of time T that a trajectory of Brownian motion stays above zero are distributed as arcsine law. This result has been extended by many researchers subsequently. Barato et al. [24] proved that the fraction of time that a thermodynamic current spends above its average value, the time a current reaches its maximum value, and the last time a current crosses its average value are also arcsine distributed. Pitman and Yor (1992) [109] extended it to a large collection of functionals derived from the lengths and signs of excursions of Brownian path away from the starting point. The long-term limiting distribution of occupation time in one-dimensional telegraph process follows the arcsine law [28]. The distribution of fiber length and stiffness in fiber-reinforced concrete (FRC) testing uses SASD-II [89].[2] Upper bounds on the approximation of the arcsine distribution

---

[2]These authors call it "cosine distribution" (p. 200), which is incorrect (see Chapter 7).

can be used to develop second-level statistical tests for detailed error analysis of the quality of pseudo-random number generators [95].

Electric conductance in rusty or disordered materials and some crystals is approximated by the arcsine law. Cyclic temperature variations of test-bed in thermal-calibration systems is assumed to be arcsine distributed with PDF $g(t; k) = k/[\pi[1 - kt^2]]$ for $-1/k < t < 1/k$, where k depends on the unit for temperature used (Celsius, Fahrenheit, etc.). It can be used to model insurance premiums, financial derivatives (economic default time of companies), optimal pricing of manufactured commodities, etc. A mixture of arcsine distributions is used to model structural credit risk in financial engineering [68]. Its applications in organic chemistry include modeling of partial melting of polymers, for which the two-parameter form is used. It is also used in chaotic dynamical systems, and in analyzing auto-regressive AR(1) processes using eigen vectors. Applications in thermodynamics can be found in Barato and Roldan (2018) [24], automobile claim frequency distribution in Kokonendji and Khoudar (2004) [87], and a discrete approximation to the arcsine distribution appears in [76]. The limiting distribution of roots of weighted orthogonal polynomials defined on $[-1, +1]$ converges to the arcsine law [45]. The lead time in some two-team games (like basketball, soccer, etc.) that a team holds a lead is approximately arcsine distributed, meaning that lead changes are most likely at the beginning and end than during the middle of the game.

5.4.1   ARCSINE TRANSFORMS

The arcsine transform is a technique similar to Fourier transform that takes a monotonic continuous function $h(x)$ as input and transforms it using the one (or more) parameter(s) version discussed above. The Fourier transform of arcsine distribution is a zero-order Bessel function of the first kind. For SASD-I with PDF

$$f(x; b) = 1/[b\pi \sqrt{(x/b)(1-x)/b}] \quad \text{for} \quad -b < x < b \tag{5.33}$$

or SASD-II with PDF

$$f(x; b) = 2/[\pi \sqrt{(b - x^2)}] \quad \text{for} \quad 0 < x < \sqrt{b} \tag{5.34}$$

we could find the arcsine transform of $h(x)$ as

$$g(b) = \int_0^\infty f(x; b)h(x)dx. \tag{5.35}$$

5.4.2   STATISTICAL PHYSICS

Statistical physics aims to study properties (symmetries, correlations, spatial dimensions, energies) of physical systems using probabilistic models of systems that interact with their environment. Macroscopic states of such systems that depend only on a few parameters (like temperature, pressure, electrical properties, etc.) are identified in the configuration space. Probability distributions are used to predict respective probabilities for the parametric space (e.g., energy landscape) and for the system to be in such a state, for which general arcsine law is a popular choice.

5.4.3   RANDOM WALKS

Several practical applications in robotics, Brownian motion, games, etc., can be formulated as random walks. For this purpose, consider a one-dimensional walk using a robot placed at the origin. It moves one step to the right with probability p, and one step to the left with probability $q = 1 - p$. Let X denotes its position after n steps. What is the distribution of X? Obviously x can take the values in the range $-n$ to $+n$ with various probabilities. The number of such walks is $N = 2^n$. Denote S by a success (move right unit step) and F by a failure (move left unit step), so that $|S - F|$ is the distance from origin (Chattamvelli and Shanmugam (2020), [42, p. 45]).

Mathematically, a random walk maps to a sequence of integers $(i_0, i_1, \ldots, i_n)$ such that $i_k - i_{k-1} = 1$ for unit-step random walks with $i_0 = 0$ if starting position is at the origin, and $i_0 = c$ (an integer) otherwise. Let $S_n = X_1 + X_2 + \cdots + X_n$ where each $X_k = \pm 1$, and $S_0 = 0$. If P_n denotes the number of positive sums ($S_n > 0$), then Levy proved that P_n/n is asymptotically arcsine distributed. Then $E(X_k) = 1 \times p - 1 \times q = p - q$, so that $E(S_n) = n(p - q)$, and $E(X_k^2) = p + q = 1$. Next, $E(S_n^2) = E(\sum_{k=1}^n X_k^2) + E(\sum_{j \neq k=1}^n X_j X_k)$. As each $X_k = \pm 1$, $X_k X_j$ takes the values -1 and $+1$ only. Hence, $E(X_k X_j) = 2pq + p^2 + q^2 = 1$. As $E(X_k^2)$ also is 1, $E(\sum_{j \neq k=1}^n X_j X_k) = 1 - 1 = 0$. Thus, $E(S_n^2) = n$. If s_n is considered as the expected distance, an interpretation of this result is that after n steps we have wandered a distance of less than $\sqrt{n}$ steps from the starting position (say origin). If the latest return to the starting point in a random walk of length $2n$ occurs when $s_{2k} = 0$, its probability of occurrence is approximately arcsine distributed with PDF $f(x; n) \sim 1/[n\pi \sqrt{x(1-x)}]$. This approaches zero as $n \to \infty$. If the epoch is taken as nx, this becomes the SASD-I because the probability of last return before epoch nx is $\sum_{k=0}^{nx}(1/n) f(k/n)$, which using $\lim_{n \to \infty} \frac{b-a}{n} \sum_{k=0}^n f(a + k(b-a)/n) \to \int_a^b f(x)dx$ becomes the CDF of SASD-I, namely $(2/\pi) \arcsin(\sqrt{x})$. Not only the latest return to origin, but the expected duration that the process spends entirely on either of the sides from the origin, and the time it reaches the minimum or maximum distance from the origin are also approximated by arcsine law SASD-I as $p(\theta) = 1/[\pi \sqrt{\theta(1-\theta)}]$, where $\theta = t/T$, with $T =$ evolving duration time counted from the start, $t =$ time duration associated with one of the events.

5.5   SUMMARY

This chapter introduced the arcsine distribution. It finds applications in random walks, Brownian motion, financial engineering and many other fields. If X_n denotes the number of steps that a random walk of length n spends on either side of the starting point, then the distribution of X_n/n approaches the arcsine law SASD-I.

CHAPTER 6

Gamma Distribution

6.1 INTRODUCTION

The gamma distribution is used to analyze skewed data, and to describe systems undergoing sequences of events or shocks that in turn lead to eventual failures called gamma-distributed degradation (GDD) models. It is popular in modeling of hydrological processes, renewal processes, and in civil engineering (performance modeling of framed structures of reinforced concrete buildings and bridges, and failure modeling of aged structures using sagging capacity loss and hogging flexural loss). The two-parameter gamma distribution can be considered as an extension of the exponential distribution. The exponential and gamma distributions (both continuous lifetime distributions) are related in a similar manner as the geometric and negative binomial distributions are in the discrete case. While the exponential distribution predicts the very first occurrence of an event (like failure of a machinery or device), the gamma law predicts the kth occurrence. Its PDF is given by

$$f(x; \lambda, m) = \lambda^m x^{m-1} e^{-\lambda x} / \Gamma(m), \quad x \geq 0, \ m > 0, \ \lambda > 0. \tag{6.1}$$

The parameter λ is called *scale parameter*, and m is the *shape parameter*. When $m = 1$, this reduces to the exponential distribution. For $m = 1/2$ we get $f(x; \lambda, m) = \sqrt{\lambda/\pi x} \ e^{-\lambda x}$. When $\lambda = 1$, the PDF becomes

$$f(x; m) = x^{m-1} e^{-x} / \Gamma(m), \quad x \geq 0, \ m > 0, \tag{6.2}$$

which for $m = 2$ is the size-biased exponential distribution. This is called the *standard form* of gamma distribution and denoted as GAMMA(m).

6.2 ALTERNATE REPRESENTATIONS

Equation (6.1) can also be written as

$$f(x; \lambda, m) = (\lambda x)^m \exp(-\lambda x) / [x \Gamma(m)], \quad x \geq 0, \ m > 0, \ \lambda > 0. \tag{6.3}$$

Replace λ by $1/\lambda$ in (6.1) to get

$$f(x; \lambda, m) = e^{-x/\lambda} x^{m-1} / [\lambda^m \Gamma(m)] = e^{-x/\lambda} (x/\lambda)^{m-1} / [\lambda \Gamma(m)]. \tag{6.4}$$

A change of scale transformation $Y = \lambda X$ (so that $dy = \lambda dx$) in (6.1) gives

$$f(y; m) = y^{m-1} e^{-y} / \Gamma(m). \tag{6.5}$$

This form is extensively tabulated. A change of origin-and-scale transformation $Y = \lambda(X - a)$ results in the three-parameter version

$$f(x; \lambda, a, m) = \exp -(x - a)/\lambda (x - a)^{m-1}/[\lambda^m \Gamma(m)], \tag{6.6}$$

where $x > a$.

6.3   RELATION TO OTHER DISTRIBUTIONS

The χ^2 distribution is a special case of gamma distribution as $\chi_n^2 = \text{GAMMA}(n/2, 1/2)$. Symbolically, if $X_1, X_2, \ldots, X_n$ are IID $Z(0, 1)$ random variables, $Y = X_1^2 + \cdots + X_n^2 \sim \text{GAMMA}(n/2, 1/2)$. If $X \sim N(\mu, \sigma^2)$, then $Y = (X - \mu)^2/(2\sigma^2)$ is a standard gamma distribution with $m = 1/2$, which is a special case of the above result. If $X_1 \sim \text{GAMMA}(a, b)$ and $X_2 \sim \text{GAMMA}(c, d)$ are independent, then $X_1/(X_1 + X_2)$ is Beta-I, and X_1/X_2 is Beta-II distributed.

Example 6.1   From independent gamma to beta   If X and Y are IID $\text{GAMMA}(\alpha, \beta_i)$, prove that $X/(X + Y)$ is Beta-I distributed.

Solution 6.2   We find the distribution of $U = X + Y$ and $V = X/(X + Y)$. The joint PDF is

$$\text{f(x,y)} = \frac{\alpha^{\beta_1 + \beta_2}}{\Gamma(\beta_1)\Gamma(\beta_2)]} x^{\beta_1 - 1} y^{\beta_2 - 1} e^{-\alpha(x+y)}. \tag{6.7}$$

The inverse mapping is $x = uv, y = u(1 - v)$, so that the Jacobian is $-u$. The joint PDF of u and v is

$$\text{f(u,v)} = \frac{\alpha^{\beta_1 + \beta_2}}{\Gamma(\beta_1)\Gamma(\beta_2)} (uv)^{\beta_1 - 1}(u - uv)^{\beta_2 - 1} e^{-\alpha u} u, \tag{6.8}$$

$0 < u < 1, 0 < v < \infty$. Combining common terms this becomes

$$\frac{\alpha^{\beta_1 + \beta_2}}{\Gamma(\beta_1)\Gamma(\beta_2)} e^{-\alpha u} u^{\beta_1 + \beta_2 - 1} v^{\beta_1 - 1}(1 - v)^{\beta_2 - 1}. \tag{6.9}$$

Integrating out u, it is easy to show that v has a $\text{Beta1}(\beta_1, \beta_2)$ distribution.

Inverse gamma distribution is obtained by a change of variable $Y = 1/X$ as

$$\text{f}(y; \lambda, m) = \lambda^m y^{-(m+1)} e^{-\lambda/y}/\Gamma(m), \quad y \geq 0, \ m > 0, \ \lambda > 0, \tag{6.10}$$

which finds applications in volume-price distribution modeling in finance. It is related to the inverse-χ^2 distribution. It is called Levy's distribution for $a = 1/2$ and $b = c/2$. Log-gamma distribution is the analogue of lognormal distribution in the Gamma case. The PDF is given by

$$f(x) = a^b/\Gamma(b)(\ln x)^{b-1} x^{-a-1}, \quad a > 1, \ b > 0. \tag{6.11}$$

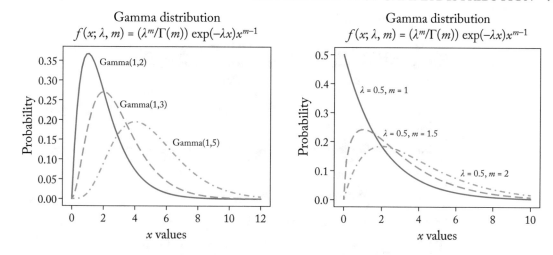

Figure 6.1: Gamma distributions.

A change of scale transformation $Y = rX$ results in GAMMA$(m, r\lambda)$ distribution. Boltzmann distribution is related to the gamma law as shown later. If the quantized energies of a molecule in an ensemble are $E_1, E_2, \ldots, E_n$, the probability that a molecule has energy E_i is given by $C \exp(-E_i/(KT))$ where K is the Boltzmann constant (gas constant divided by Avogadro number) and T is the absolute temperature. The sum of the energies is gamma distributed when $E_i's$ are independent. Its relation with Poisson distribution is discussed below. Other relationships can be found in Song (2005) [137].

6.4   PROPERTIES OF GAMMA DISTRIBUTION

There are two parameters, both of which are real numbers. As mentioned above, for $\lambda = 1$, we get the one-parameter standard gamma distribution with PDF

$$f(x;m) = x^{m-1}e^{-x}/\Gamma(m), \quad \text{for } x > 0. \tag{6.12}$$

This is called Erlang distribution ERL(m) when m is an integer, which represents the waiting time for m independent events to occur in which the time of each event occurrence has an exponential distribution with mean n. The coefficient of skewness and kurtosis are $1/\sqrt{m}$ and $3(1 + 2/m)$ which are both independent of λ. This distribution is always leptokurtic (see Figure 6.1). See Table 6.1 for further properties.

6.4.1   ADDITIVITY PROPERTY

This distribution can be obtained as the sum of m IID exponential variates with parameter λ, resulting in GAMMA(m, λ). If X and Y are two IID gamma random variables with the

Table 6.1: Properties of gamma distribution ($\lambda^m x^{m-1} e^{-\lambda x} / \Gamma(m)$)

Property	Expression	Comments
Range of X	$x \geq 0$	Continuous
Mean	$\mu = m/\lambda$	
Median	$\log(2)/\lambda$	
Mode	$(m-1)/\lambda$	
Variance	$\sigma^2 = m/\lambda^2$	$\sigma^2 = \mu^2/m = \mu/\lambda$
Skewness	$\gamma_1 = 2/\sqrt{m}$	
Kurtosis	$\beta_2 = 3/(1 + 2/m)$	Always leptokurtic
MD	$2m^m e^{-m}/(\lambda \Gamma(m))$	$2\lambda^m/\Gamma(m) \int_0^{m/\lambda} \gamma(x; \lambda, m) dx$
CV	$1/\sqrt{m}$	
CDF	$P(x; \lambda, m)$	$\frac{\lambda^m}{\Gamma(m)} \int_0^x e^{-\lambda y} y^{m-1} dy$
Moments	$\mu'_r = \Gamma(k + m)/\lambda^{k+m}$	
MGF	$[\lambda/(\lambda - t)]^m$	
ChF	$[\lambda/(\lambda - it)]^m$	

Replace λ by $1/\lambda$ for the alternate parametrization. $\sigma^2 > \mu^2$ when $m < 1$. $\frac{\lambda^m}{\Gamma(m)} \int_0^x e^{-\lambda y} y^{m-1} dy$ is called the incomplete gamma function.

same scale parameter λ, and shape parameters m_1 and m_2, respectively, their sum $X + Y$ is distributed as gamma with the same scale parameter and $m_1 + m_2$ as shape parameter. This result can be generalized to any number of independent gamma variates as "the sum of m independent gamma variates with shape parameters m_i and the same scale parameter λ is distributed as GAMMA($\sum_i m_i, \lambda$)." It follows from this result that the gamma distribution is infinitely divisible.

Problem 6.3   If $X_1 \sim$ GAMMA(c,b) and $X_2 \sim$ GAMMA(c, d) are independent, then prove that $(X_1 + X_2)$ and X_1/X_2 are independently distributed.

6.4.2   MOMENTS AND GENERATING FUNCTIONS

The raw moments are easy to find using gamma integral. Consider

$$E(X^k) = \int_0^\infty \lambda^m x^k x^{m-1} e^{-\lambda x} / \Gamma(m). \qquad (6.13)$$

Using gamma integral this becomes $\Gamma(k+m)/\lambda^{k+m}$. From this we get the mean as $\mu = m/\lambda$ and variance $\sigma^2 = m/\lambda^2 = \mu/\lambda$. This shows that the variance is more than the mean for $\lambda < 1$, and *vice versa*. The coefficient of variation (CV) of gamma distribution is $(\mu/\lambda)/\mu = 1/\lambda$, which is used to understand reaction kinetics in chemical engineering (the particle shape deviates from spherical toward flat as $\lambda \to 0$). The characteristic function is

$$\phi(t) = (\lambda^m/\Gamma(m)) \int_0^\infty x^{m-1} e^{-x(\lambda-it)} dx. \tag{6.14}$$

Put $Y = (\lambda - it)X$, so that $dy = (\lambda - it)dx$. The range of integration remains the same, and we get

$$\phi(t) = (\lambda^m/[\Gamma(m)(\lambda - it)^m]) \int_0^\infty y^{m-1} e^{-y} dy = (\lambda/(\lambda - it))^m = (1 - it/\lambda)^{-m}, \tag{6.15}$$

for $t < \lambda$. By expanding this as an infinite series we get $(1 - it/\lambda)^{-m} =$

$$\sum_{k=0}^\infty \binom{m+k-1}{k} (-it/\lambda)^k = 1 + (m/\lambda)it + m(m+1)/[1*2](it/\lambda)^2 + \cdots \tag{6.16}$$

Problem 6.4 If X is distributed as $\Gamma(\alpha, 1)$, find the distribution of $Y = \log(X/\alpha)$.

6.5   TAIL AREAS

Left-tail area of gamma distribution is called incomplete gamma function ratio, and the integral $\gamma(\lambda, m, x) = \int_0^x e^{-\lambda y} y^{m-1} dy$ (without the constants) as the incomplete gamma function. It is called complete gamma function when the upper limit of integration is ∞. The CDF of gamma distribution is a sum of Poisson probabilities when the shape parameter p is an integer:

$$F(x; m, p) = 1 - \sum_{k=0}^{p-1} e^{-\lambda x} (\lambda x)^k / k!. \tag{6.17}$$

The corresponding right-tail area is denoted as $S(x; m, p) = \frac{m^p}{\Gamma(p)} \int_x^\infty e^{-mt} t^{p-1} dt$. Both of these are extensively tabulated. For example, Pearson tabulated the function $I(x, p) = \frac{1}{\Gamma(p+1)} \int_0^{x\sqrt{p+1}} e^{-t} t^p dt$. It has a representation in terms of confluent hypergeometric functions as

$$\gamma(x; 1, p) = (x^p/p)e^{-x} {}_1F_1(1, p+1; x) = (x^p/p) {}_1F_1(p, p+1; -x), \tag{6.18}$$

and error function as $\gamma(x^2; 1, 1/2) = \mathrm{erf}(x)$.

6.5.1   INCOMPLETE GAMMA FUNCTION (IGF)

The function $\Gamma(m) = \int_0^\infty x^{m-1} e^{-x} dx$ is called the complete gamma function. When m is an integer, the above integral becomes $\Gamma(m) = (m-1)!$. When m is not an integer, we get the recurrence $\Gamma(m) = (m-1) * \Gamma(m-1)$ using integration by parts.

Definition 6.5   The left-tail area of gamma distribution is called the incomplete gamma function. It is given by

$$P(x; \lambda, m) = \frac{\lambda^m}{\Gamma(m)} \int_0^x e^{-\lambda y} y^{m-1} dy. \tag{6.19}$$

As there are two parameters, a simple change of scale transformation mentioned above can be made. This gives one-parameter gamma distribution. The integral with and without the normalizing constant is denoted as $\gamma(x; m) =$

$$\int_0^x y^{m-1} e^{-y} dy = \Gamma(m) - \Gamma(x; m) \quad \text{and} \quad P(x, m) = \frac{1}{\Gamma(m)} \int_0^x y^{m-1} e^{-y} dy. \tag{6.20}$$

These satisfy the recurrence

$$\gamma(x; m+1) = m * \gamma(x; m) - x^m e^{-x} \quad \text{and} \quad P(x, m+1) = P(x; m) - \frac{x^m e^{-x}}{\Gamma(m+1)}.$$

Put $y = x^2$ and $m = 1$ to get $\gamma(1, x^2) = \int_0^{\sqrt{y}} e^{-y^2} dy$. An approximate relation with erfc() is available as $\gamma(x, m)/\Gamma(m) \sim 0.5 * \text{erfc}(-K\sqrt{m/2})$ where $K^2 = 2(x/m - 1 - \ln(x/m))$.

Example 6.6   Mean deviation of gamma distribution   Find the mean deviation of the gamma distribution.

Solution 6.7   We have seen above that the CDF is $P(x; \lambda, m)$. As the mean is $\mu = m/\lambda$ we get the MD using Theorem 1.12 in page 10 as

$$MD = 2 \int_{ll}^{\mu} F(x) dx = 2 \int_0^{m/\lambda} P(x; \lambda, m) dx = 2m^m \exp(-m)/[\lambda \Gamma(m)]. \tag{6.21}$$

Example 6.8   Sum and ratio of independent gamma distributions   If X and Y are IID GAMMA(α, β_i), find the distribution of (i) $X + Y$, (ii) X/Y.

Solution 6.9   Let $U = X + Y$ and $V = X/Y$. Solving for X and Y in terms of U and V, we get $x = \frac{uv}{1+v}$, and $y = \frac{u}{1+v}$. The Jacobian of the transformation is

$$|J| = \begin{vmatrix} \frac{v}{1+v} & \frac{u}{(1+v)^2} \\ \frac{1}{1+v} & -\frac{u}{(1+v)^2} \end{vmatrix} = \frac{-u}{(1+v)^2}.$$

The joint PDF of X and Y is $f(x,y) = \frac{\alpha^{\beta_1+\beta_2}}{\Gamma(\beta_1)\Gamma(\beta_2)} e^{-\alpha(x+y)} x^{\beta_1-1} y^{\beta_2-1}$. Multiply by the Jacobian, and substitute for x, y to get

$$f(u,v) = \frac{\alpha^{\beta_1+\beta_2}}{\Gamma(\beta_1)\Gamma(\beta_2)} e^{-\alpha u} (uv/(1+v))^{\beta_1-1} (u/(1+v))^{\beta_2-1} \frac{u}{(1+v)^2}. \tag{6.22}$$

The PDF of u is obtained by integrating out v as

$$f(u) = \frac{\alpha^{\beta_1+\beta_2}}{\Gamma(\beta_1)\Gamma(\beta_2)} e^{-\alpha u} u^{\beta_1+\beta_2-1} \int_0^\infty v^{\beta_1-1}/(1+v)^{\beta_1+\beta_2} dv. \tag{6.23}$$

Put $1/(1+v) = t$ so that $v = (1-t)/t$, and $dv = -1/t^2 dt$. This gives us

$$\begin{aligned} f(u) &= \frac{\alpha^{\beta_1+\beta_2}}{\Gamma(\beta_1)\Gamma(\beta_2)} e^{-\alpha u} u^{\beta_1+\beta_2-1} \int_0^1 t^{\beta_2-1} (1-t)^{\beta_1-1} dt \\ &= \frac{\alpha^{\beta_1+\beta_2}}{\Gamma(\beta_1)\Gamma(\beta_2)} e^{-\alpha u} u^{\beta_1+\beta_2-1} B(\beta_1, \beta_2). \end{aligned} \tag{6.24}$$

This simplifies to

$$f(u) = \frac{\alpha^{\beta_1+\beta_2}}{\Gamma(\beta_1+\beta_2)} e^{-\alpha u} u^{\beta_1+\beta_2-1}, \tag{6.25}$$

which is GAMMA(β_1, β_2).

The PDF of v is found by integrating out u as

$$f(v) = \frac{\alpha^{\beta_1+\beta_2}}{\Gamma(\beta_1)\Gamma(\beta_2)} \frac{v^{\beta_1-1}}{(1+v)^{\beta_1+\beta_2}} \int_{u=0}^\infty u^{\beta_1+\beta_2-1} e^{-\alpha u} du. \tag{6.26}$$

This simplifies to $f(v) = \frac{\Gamma(\beta_1+\beta_2)}{\Gamma(\beta_1)\Gamma(\beta_2)} \frac{v^{\beta_1-1}}{(1+v)^{\beta_1+\beta_2}}$, which is Beta2$(\beta_1, \beta_2)$ (also called Pearson type VI distribution).

Example 6.10   A critical component in a system has a lifetime distribution EXP(λ), and it is replaced with an identical component immediately after failure, so that the downtime is negligible. What is the number of spares to be kept in inventory if the system must operate n hours continuously with probability p (say p $\geq$ .99)?

Solution 6.11   Let X denote the lifetime of a single component and m denote the number of spares to be kept. Then $Y = X_1 + X_2 + \cdots + X_m$ denotes the entire lifetime (excluding the component already inside the system).[1] As the components are independent with the same lifetime, Y has a gamma distribution GAMMA(m, λ). Thus, we have to find n such that $\Pr[Y \geq n] = p$. This can be solved either using tables of gamma distribution or iteratively.

Problem 6.12   If a positive random variable has a conditional PDF $f(x; r, \lambda) = x^{r-1} \lambda^r \exp(-\lambda x)/\Gamma(r)$, and the system's parameter $\lambda > 0$ has an unconditional PDF

---

[1]if the component already inside is also counted, then $Y = X_1 + X_2 + \cdots + X_{m+1} \sim$ GAMMA$(m+1, \lambda)$.

$f(\lambda; s, \mu) = \lambda^{s-1} \mu^s \exp(-\mu\lambda)/\Gamma(s), \quad s > 0, \mu > 0$, then prove that the random variable X has an unconditional marginal PDF $f(x; r, s, \mu) = x^{r-1} \mu^s \Gamma(r + s)/[(\mu + x)^{r+s} \Gamma(r)\Gamma(s)]$.

6.5.2   FITTING

If $\hat{\lambda}$ and $\hat{m}$ denote the MoM estimates of the parameters λ and m in (6.1), they are related as $\hat{m}/\hat{\lambda} = \bar{x}$ and $\hat{m}/\hat{\lambda}^2 = m_2$, where $m_2 = (1/n) \sum_{k=1}^{n} (x_k - \bar{x})^2$. From these, $\hat{\lambda} = \bar{x}/m_2$ and $\hat{m} = \bar{x}^2/m_2$. As the gamma distribution can take a variety of shapes, these estimators are often biased. Modified method of moments estimates are more accurate (Cohen and Whitten (1986) [44]). The log-likelihood function is

$$\log(L(x; \hat{\lambda}, m)) = (m - 1) \sum_{k} \log(x_k) - n \log(\Gamma(m)) + nm \log(\lambda) - \lambda \sum_{k} x_k. \tag{6.27}$$

As λ is present only in the last two terms, an estimate of it is easily obtained by differentiating w.r.t. λ and equating it to zero, to get $\hat{\lambda} = \hat{m}/\bar{x}$. Substitute this value of $\hat{\lambda}$ in (6.27) to get

$$\log(L(x; \lambda, m)) = (m - 1) \sum_{k} \log(x_k) - n \log(\Gamma(m)) + nm \log(m) - nm \log(\bar{x}) - nm. \tag{6.28}$$

As $m \log(m)$ is a convex function, we could iteratively maximize a lower bound to obtain the MLE of m. Start with an initial value m_0. Then the maximum is given by $n[(1 + \log(m_0)) - 1 - \log(\bar{x}) + \overline{\log(\bar{x})}] = \psi(\hat{m})$ where $\psi(x) = \partial/\partial x [\log \Gamma(x)]$ is the di-gamma function, and $\overline{\log(\bar{x})} = (1/n) \sum_{k} \log(x_k) = (1/n) \log(x_1 x_2 \ldots x_n) = \log((x_1 x_2 \ldots x_n)^{1/n}) = \log(GM)$, so that $\log(\hat{m}) = \psi(m) + \log(\bar{x}/GM)$ where GM $= (\prod_{k=1}^{n} x_k)^{1/n}$ is the geometric mean.

6.6   GENERAL GAMMA DISTRIBUTIONS

Generalized gamma distribution finds applications in a variety of fields like speech signal processing, rainfall and flood modeling, particle size modeling in metallurgy, etc. The exponentiated gamma distribution (EGD) has PDF

$$f(x; \lambda, \theta) = (1 - \exp(-\lambda x)(1 + \lambda x))^{\theta} \quad \text{for} \quad \lambda, \theta > 0, \quad \text{and} \quad x \geq 0. \tag{6.29}$$

A three-parameter generalized gamma distribution (GGD) of the form

$$f(x; a, b, c) = cb^{a/c} x^{a-1} \exp(-bx^c)/\Gamma(a/c), \tag{6.30}$$

is used in luminosity modeling in astrophysics (galactic red-shifts) ([148], [149]). This reduces to the gamma law for $c = 1$, and to the exponential law for $a = c = 1$. The CDF is obtained by integration as $F(x; a, b, c) = 1 - \Gamma(a/c, bx^c)/\Gamma(a/c)$. This has mean $\mu = \Gamma((a + 1)/c)/[\Gamma(a/c)b^{1/c}]$ and mode $[(a - 1)/(bc)]^{1/c}$. The kth moment is $\mu'_k = \Gamma((a + k)/c)/[\Gamma(a/c)b^{k/c}]$.

By reflecting the three-parameter gamma distribution with PDF

$$f(x; a.b.c) = (x - a)^{c-1} \exp(-(x - a)/b)/[b^c \Gamma(c)] \tag{6.31}$$

for $x \geq a$ we get the extended Laplace distribution (also called reflected gamma distribution) has PDF

$$f(x; a, b, c) = |x - a|^{c-1} \exp(-|x - a|/b)/[2b^c \Gamma(c)] \tag{6.32}$$

for $-\infty < x < \infty$, and $b > 0$. This was extended by Küchler and Tappe (2008) [91] to model financial market fluctuations as

$$f(x; a, b, c) = (|\lambda|^\alpha / \Gamma(\alpha)) |x|^{\alpha-1} \exp(-\lambda x), \quad \text{for } x \in \mathbb{R} - 0, \tag{6.33}$$

which is the regular gamma distribution for $\lambda > 0$, and the reflected gamma distribution for $\lambda < 0$. This has characteristic function $\phi(t) = (\lambda/(\lambda - it))^\alpha$.

6.6.1   TRUNCATED GAMMA DISTRIBUTIONS

Left-truncated gamma distribution is obtained by truncating at a positive integer c. The resulting PDF is

$$f(x; m, \lambda, c) = \lambda^m x^{m-1} e^{-\lambda x}/[\Gamma(m)\Gamma(m, c)], \tag{6.34}$$

where $\Gamma(m, c) = 1 - \gamma(m, c)$ is the SF of gamma distribution. Similarly, it can be truncated at upper tail or both tails.

Problem 6.13   The radial wave function at ground state of the hydrogen atom is given by $f(z) \propto z^{-3/2} \exp(-rz)$ for $z \geq 0$. Find the unknown constant and mean value.

6.7   APPLICATIONS

Gamma distribution is used to model the distribution of high wealth-range (using respective currencies as units) among members of a society or a country (Pareto distribution is used for the low and medium wealth ranges). Wind-energy modeling applications use a variant of gamma distribution called the general Burr-Gamma distribution (GBGD(μ, σ, k, t) [27] where μ is the location, σ is the scale, k is the skewness, and t is the tail parameter; and μ itself is a linear model built using quadrant-wise presence or absence of wind, the wind direction, etc.), and one-parameter gamma distribution GAMMA($m, 1$) for goodness-of-fit tests [51]. Chemical engineers use it to model particle size distributions (the other competitive size distributions being the Rosin–Rammler–Bennett (RRB) distribution and Gate–Gaudin–Schuhmann (GGS) distribution) when an assemblage of particles of different shapes are involved, and in exponential orbital modeling in quantum chemistry. Electronics engineers use it to approximate the radial wave functions of atoms (like Hydrogen) at different electrostatic potentials (like $1s$ (spherically symmetric), $2p$ (directed along z axis), etc.) which represent the scale parameter λ.

The three-parameter gamma distribution is used to model the fraction of charge between spherical shells in ionic atmospheres as

$$f(r; a, k, q_j) = (q_j k^2 / (1 + ka)) \exp(-k(r - a)), \quad \text{for} \quad r > a, \tag{6.35}$$

where q_j is the electrostatic potential acting on the jth ion at location r, a is the radius of all ions, and k is the ionic stretch, which depends on the enclosing temperature. It is used in mining engineering to model particle sizes in coal flotation kinetics.

Whitson et al. (2007) used the three-parameter gamma distribution to true boiling point (TBP) data of stabilized petroleum liquid (stock tank oil and condensate), and for describing the molar distributions of heptanes-plus (C7+) fractions. Civil engineers use it to model total deterioration of structural resistance of concrete structures and other civil infrastructures over time t as

$$f(x; a(t), b) = [b^{a(t)} / \Gamma(a(t))] \, x^{a(t)-1} \exp(-bx), \tag{6.36}$$

where the shape-parameter $a(t)$ is a function of time. Incremental deterioration between times t_1 and t_2 is assumed to be independent, and follows the gamma process[2] with $X(0) = 0$. The distribution of $X(t_2) - X(t_1)$ is then modeled using gamma law [141]. Meteorologists use the lognormal law for low-rainfall, and the gamma law for high rainfall (amounts of daily rainfall at a fixed locality) and for the ascension curve of the hydrograph modeling. The two-parameter gamma distribution is used as a conjugate prior in Bayesian inference, and in petroleum engineering (optimal capacity and deterioration modeling of oil and gas pipelines). Truncated gamma distributions find applications in astronomy and ultraviolet photoelectronics. Luminosity of astrophysical objects (like galaxies and quasars) are approximated using generalized and truncated gamma distributions.

6.7.1   METALLURGY

Quantitative modeling of mineral processing operations use multi-parameter statistical distributions. Physical characteristics of individual particles like its size and mineralogical composition that vary when subjected to treatment in an ore dressing equipment are the most important in metallurgy. Other less important properties include brittleness, specific gravity, deviations from regular shapes (cubes, sphere, etc.). When a parent particle of size D is broken into smaller ones, the progeny particles are modeled using right-truncated distributions with D as the upper limit. Mineral content and fluid-solid kinetics are modeled using statistical distributions with at least four parameters—average (μ), dispersion (σ), mass fractions of particles that contain only a single mineral. Prime candidates include general beta and gamma distributions. The induction time (which is the time taken from the instant of direct collision to a temporary stable

---

[2]Gamma process is a stochastic process in which independent (continuous) increments follow a gamma law with time dependent shape parameter, and either a constant or a linear function of design variables (e.g., a linear regression model) as scale parameter.

resting position by three-phase contact) is a function of particle size, contact angle and chemical conditioning of the particle also is assumed to follow the two-parameter gamma law.

6.7.2   RELIABILITY

Time-dependent reliability analysis is critical in civil engineering to model age-related deterioration, and to predict optimal life-spans and replacement policies. If the deterioration is gamma-distributed, the lifetime is inverse gamma-distributed. There exist multiple types of uncertainties in structural reliability. The prominent among them are temporal reliability, sampling (site-specific) reliability, reliability against infrequent events (unusual events like fires, earthquakes, floods, extreme climate, etc). Structural analysis uses statistical distributions (e.g., gamma, normal, lognormal, and Pareto distributions)[3] to build probabilistic prediction models in which precarious parameters are functions of time (from the commissioning of the structure). This applies to entire structure in temporal reliability modeling, and to sub-parts thereof in other models.

The random variable approach is the preferred choice in temporal reliability models and the random process approach in other models. An advantage of random variable deterioration models is that critical parameters can be randomized within permissible levels using multiple statistical distributions. The random process model has an advantage in modeling cumulative damage in time interval (t_1, t_2) conditional on accumulated damages and repair work until time t_1 by considering uncertain damages as a stochastic process. Loss of structural capacity (for static structures) and of projected performance (for dynamic structures like conveyor belts, escalators) assumes importance when failures can result in injury or human lives. Strength of materials that resulted in the failure (called yield in structural engineering, which is used in yield-line graphs) are obtained and analyzed to get more insight on optimal loads (safety margins) or in extending life spans. Empirical models built from yield data can be used to schedule inspections and repair work.

6.8   SUMMARY

The gamma distribution is a popular choice in many engineering and applied scientific fields like chemical, civil and petroleum engineering, metallurgy, wind-energy modeling, reliability, etc. This chapter introduced the gamma distributions, and their basic properties. It also discussed tail areas of gamma distribution using incomplete gamma function. Special gamma distributions like truncated and log-gamma distributions are briefly introduced.

---

[3]Uncertainties are modeled using the Gaussian law, while wear and tear, corrosion, erosion, etc. that accumulate over time are modeled using gamma law. In the case of discrete deterioration events, shocks are assumed to be gamma distributed and occur at small intervals.

C H A P T E R 7

Cosine Distribution

7.1 INTRODUCTION

The cosine distribution is used to model the flow of compressed fluids and gases at a molecular level, in beam-scattering studies of surface-chemistry, for optimal performance of vacuum operated machinery, and in the design of optimal bearing loads of cylinders and rings. It also finds applications in moving robots and self-driving autonomous vehicles for avoiding disastrous collisions with obstacles on their way. The PDF of raised cosine distribution—type I (RCD-I) is given by

$$f(x; a, b) = 1/(2b) \ \cos((x - a)/b), \quad a - b\pi/2 \leq x \leq a + b\pi/2, \ b > 0. \tag{7.1}$$

To show that this is indeed a PDF, integrate over $[a - b\pi/2, a + b\pi/2]$ to get

$$\int_{a-b\pi/2}^{a+b\pi/2} f(x)dx = \frac{1}{2b} \int_{a-b\pi/2}^{a+b\pi/2} \cos((x - a)/b)dx. \tag{7.2}$$

Put $u = (x - a)/b$, so that $du = dx/b$. The limits become $-\pi/2$ to $\pi/2$, so that

$$\int_{a-b\pi/2}^{a+b\pi/2} f(x)dx = \frac{1}{2} \int_{-\pi/2}^{\pi/2} \cos(u)du = \frac{1}{2} \sin(u)|_{-\pi/2}^{\pi/2} = 1. \tag{7.3}$$

Put $a = 0$ in (7.1) to get origin-centered cosine distribution

$$f(x; b) = \cos(x/b)/(2b), \quad -b\pi/2 \leq x \leq b\pi/2, \ b > 0. \tag{7.4}$$

Next, put $a = 0, b = 1$ in (7.1) to get the standard cosine distribution (SRCD-I)

$$f(x) = (1/2) \ \cos(x) = 0.5 \cos(x), \quad -\pi/2 \leq x \leq \pi/2. \tag{7.5}$$

7.2 ALTERNATE REPRESENTATIONS

As a is the location, and b is the scale parameter, the notation $\mu = a$ and $\sigma = b$ is sometimes used as

$$f(x; \mu, \sigma) = (1/(2\sigma)) \ \cos((x - \mu)/\sigma), \quad \mu - \sigma\pi/2 \leq x \leq \mu + \sigma\pi/2, \ \sigma > 0. \tag{7.6}$$

Another representation in which the range is $[a - b, a + b]$ has PDF

$$f(x; a, b) = (\pi/(4b)) \ \cos(\pi(x - a)/(2b)), \quad a - b \leq x \leq a + b, \ b > 0. \tag{7.7}$$

To prove that this is indeed a PDF, make the transformation $u = \pi(x - a)/(2b)$ so that $2bdu = \pi dx$ (alternately, put $u = (x - a)/(2b)$ and use $\int \cos(\pi u)du = \sin(\pi u)/\pi$). The limits of integration become $[-\pi/2, \pi/2]$, and we get

$$(\pi/(4b))((2b)/\pi) \int_{-\pi/2}^{\pi/2} \cos(u)du = (1/2)\sin(u)|_{-\pi/2}^{\pi/2} = 1. \tag{7.8}$$

Put $a = 0$ to get the half-cosine distribution as

$$f(x;b) = (\pi/(4b))\cos(\pi x/(2b)), \quad -b \leq x \leq b, \ b > 0. \tag{7.9}$$

Another representation called raised cosine distribution—type II (RCD-II) has PDF

$$f(x;a,b) = (1/(2b))[1 + \cos(\pi(x - a)/b)], \quad a - b \leq x \leq a + b, \ b > 0. \tag{7.10}$$

Integrate over the range, split into two integrals, and use $\sin(-\pi) = \sin(\pi) = 0$ in the second one to prove that this is indeed a PDF. This has variance $b^2(1/3 - 2/\pi^2)$ (see page 84). The PDF reduces to

$$f(x;b) = (1/(2b)) [1 + \cos(\pi x/b)], \quad -b \leq x \leq b, \ b > 0, \tag{7.11}$$

for $a = 0$, which is used in error distribution modeling. Put $a = 0, b = 1$ in (7.10) to get the standard raised cosine distribution (SRCD-II)

$$f(x) = [1 + \cos(\pi x)]/2, \quad -1 \leq x \leq 1. \tag{7.12}$$

Changing the scale as $Y = \pi X$ results in the alternate form

$$f(x) = [1 + \cos(x)]/(2\pi), \quad -\pi \leq x \leq \pi. \tag{7.13}$$

The domain in (7.1) can also be extended to other ranges by scaling as follows:

$$f(x;a,b) = (1/(4b)) \cos((x - a)/(2b)), \quad a - b\pi \leq x \leq a + b\pi, \ b > 0. \tag{7.14}$$

The x is usually an "angle direction" (like wind-speed direction, location of an object w.r.t. the normal in radar), angular scattering of molecules in motion around a specularly reflected incident particle direction, or angles of cornea curvature (Shanmugam (2020) [129]). It can also be an angle in a transformed space as in automatic color recognition problems using chroma of the color in RGB coordinate systems. A close analogue of the cosine distribution is the *cardioid distribution* with PDF

$$f(x;a,b) = 1/(2\pi) [1 - 2b \cos(x - a)], \quad 0 \leq x \leq 2\pi, \ 0 < b < 0.5, \tag{7.15}$$

that has applications in electronics engineering and other fields. This has CDF $F(x;a,b) = \frac{1}{\pi}(x/2 - b\sin(x - a))$. This reduces to the cosine distribution for $b = -1/2$. Random samples are generated most easily from those of CUNI(-1,1) distribution using the relationship $x = a + b\sin^{-1} u$.

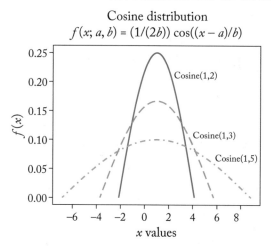

Figure 7.1: Cosine distributions.

7.3 PROPERTIES OF COSINE DISTRIBUTION

As this distribution is symmetric, the mean, median, and mode coincide at $x = a$, which is the location parameter; all odd moments (except the first) are zeros, and skewness is also zero (Figure 7.1). It is a bounded distribution (finite-support), which is the preferred choice for error modeling (other choices being truncated Gaussian and arcsin distributions).

7.3.1 MEAN AND VARIANCE

The mean of $f(x; a, b) = 1/(2b) \cos((x - a)/b)$ is given by

$$\mu = E(x) = \frac{1}{2b} \int_{a-\pi b/2}^{a+\pi b/2} x \cos((x - a)/b) dx. \tag{7.16}$$

Put $u = (x - a)/b$ (so that $dx = b\, du$) to get

$$\mu = \frac{1}{2b} \int_{-\pi/2}^{\pi/2} (a + bu) \cos(u) b\, du$$
$$= \frac{1}{2} \left(a \int_{-\pi/2}^{\pi/2} \cos(u) du + b \int_{-\pi/2}^{\pi/2} u \cos(u) du \right) = I_1 + I_2. \tag{7.17}$$

As I_2 is the integral of an odd function, it evaluates to zero; and I_1 evaluates to $\frac{a}{2} \sin(u)|_{-\pi/2}^{\pi/2} = a$. Thus, we get $\mu = a$. Now consider

$$E(x^2) = \frac{1}{2b} \int_{a-\pi b/2}^{a+\pi b/2} x^2 \cos((x - a)/b) dx. \tag{7.18}$$

Put $u = (x - a)/b$ $(dx = bdu)$ to get

$$E(x^2) = \frac{1}{2b} \int_{-\pi/2}^{\pi/2} (a + bu)^2 \cos(u)\, b\, du$$

$$= \frac{1}{2} \int_{-\pi/2}^{\pi/2} (a^2 + 2abu + b^2u^2) \cos(u) du = I_1 + I_2 + I_3. \tag{7.19}$$

Now, $I_1 = (a^2/2) \int_{-\pi/2}^{\pi/2} \cos(u) du = a^2$, and I_2 being an odd function vanishes. As I_3 is an even function, we get $I_3 = \frac{b^2}{2} \times 2 \int_0^{\pi/2} u^2 \cos(u) du = b^2 \left[-2\sin(u) + 2u\cos(u) + u^2 \sin(u)|_{-\pi/2}^{\pi/2} \right]$ $= b^2(\pi^2/4 - 2)$. From this $Var(X) = \sigma^2 = a^2 + b^2(\pi^2/4 - 2) - a^2 = b^2(\pi^2/4 - 2)$.

To find the variance of the alternate representation in (7.7), proceed as above by finding $E(X^2) = (1/(2b)) \int_{a-b}^{a+b} x^2 [1 + \cos(\pi(x-a)/b)] dx$. Split into two integrals as $I_1 + I_2$ where $I_1 = (1/(2b)) \int_{a-b}^{a+b} x^2 dx = (1/(2b)) x^3/3 |_{a-b}^{a+b} = [(a+b)^3 - (a-b)^3]/(6b)$. Expand the cubic and cancel out common factors to get this as $I_1 = a^2 + b^2/3$. Next, $I_2 = 1/(2b) \int_{a-b}^{a+b} x^2 \cos(\pi(x-a)/b) dx$. Put $u = \pi(x-a)/b$ $(dx = (b/\pi)du)$ to get $I_2 = (1/(2b))(b/\pi) \int_{-\pi}^{\pi} (a + bu/\pi)^2 \cos(u) du$. Expand the quadratic term and split this into three integrals $I_3 + I_4 + I_5$, where $I_3 = \int_{-\pi}^{\pi} (a^2/(2\pi)) \cos(u)\, du = a^2/(2\pi) \sin(u)|_{-\pi}^{\pi} = 0$. As I_4 contains $u\cos(u)$ (an odd function) it evaluates to zero. Next consider $I_5 = (1/(2\pi))(b^2/\pi^2) \int_{-\pi}^{\pi} u^2 \cos(u) du = (b^2/(2\pi^3))(-2\sin(u) + 2u\cos(u) + u^2 \sin(u))|_{-\pi}^{\pi}$. The first and third terms vanish as $\sin(\pi) = 0$, and $I_5 = (b^2/(2\pi^3))(-4\pi) = -2b^2/\pi^2$. Thus, $I_3 + I_4 + I_5 = -2b^2/\pi^2$, from which $I_1 + I_2 = a^2 + b^2/3 - 2b^2/\pi^2$. From this $\sigma^2 = E(X^2) - [E(x)]^2 = b^2/3 - 2b^2/\pi^2 = (b^2/3)(1 - 6/\pi^2) = b^2(1/3 - 2/\pi^2)$, which depends only on b. The even moments of $f(x; a, b) = \frac{1}{2b}[1 + \cos(\pi(x-a)/b)]$ are given by

$$E(X^{2k}) = \frac{1}{2} \int_{-1}^{1} x^{2k} [1 + \cos(\pi x)] dx = \frac{1}{n+1} + \frac{1}{2n+1}\, {}_1F_2\left(n + \frac{1}{2}; \frac{1}{2}, n + \frac{3}{2}; -\pi^2/4\right), \tag{7.20}$$

where ${}_1F_2()$ is the generalized hypergeometric function.

7.3.2   TAIL AREAS

The CDF of $f(x; a, b) = \frac{1}{2b} \cos(x-a)/b$ is given by

$$F(x) = \int_{a-\pi b/2}^{x} \frac{1}{2b} \cos((y-a)/b) dy. \tag{7.21}$$

As before, put $u = (y-a)/b$ to get

$$F(x) = \int_{-\pi/2}^{(x-a)/b} \frac{1}{2} \cos(u) du = \frac{1}{2} \sin(u)|_{-\pi/2}^{(x-a)/b} = \frac{1}{2} [\sin((x-a)/b) + 1]. \tag{7.22}$$

Table 7.1: Properties of cosine distribution ($\frac{1}{2b}\cos((x-a)/b)$)

Property	Expression	Comments
Range of X	$a - b\pi 2b \leq x \leq a + b\pi/2$	Continuous; finite
Mean	$\mu = a$	Median = mode = a
Variance	$\sigma^2 = (\pi^2 - 8)b^2$	
Skewness	$\gamma_1 = 0$	Symmetric
Kurtosis	$\beta_2 = 9$	Always leptokurtic
MD	$b(\pi - 2)$	$Q_1 = a - b\pi/6, Q_3 = a + b\pi/6$
CDF	$\frac{1}{2}[1 + \sin((x-a)/b)]$	

The CDF of $f(x; a, b) = \frac{\pi}{2b}\cos(\pi(x-a)/b)$ follows similarly as

$$F(x; a, b) = \frac{\pi}{2b}\int_{a-b}^{x}\cos(\pi(y-a)/b)dy = \frac{1}{2}[\sin(\pi(x-a)/b) + 1]. \qquad (7.23)$$

The CDF of (7.10) is

$$F(x; a, b) = \frac{1}{2}[1 + (x-a)/b + \frac{1}{\pi}\sin(\pi(x-a)/b)], \qquad (7.24)$$

and the characteristic function is $\phi(t; a, b) = \pi^2 \sin(bt)/[bt(\pi^2 - b^2 t^2)]\exp(iat)$. The CDF of $f(x) = [1 + \cos(x)]/(2\pi), \ -\pi \leq x \leq \pi$ in (7.13) is

$$F(x) = [\pi + x + \sin(x)]/(2\pi), \quad -\pi \leq x \leq \pi. \qquad (7.25)$$

See Table 7.1 for further properties.

7.4   TRUNCATED COSINE DISTRIBUTIONS

Truncated cosine distributions are obtained by truncating in left, right or both tails. Consider $f(x; a, b) = 1/(2b)\cos((x-a)/b), \ a - b\pi/2 \leq x \leq a + b\pi/2, \ b > 0$. If it is truncated in the left tail at c, the PDF becomes

$$f(x; a, b, c) = 1/[(2b)(1 - \frac{1}{2}[\sin((c-a)/b) + 1])] \cos((x-a)/b)$$
$$= 1/[b(1 - [\sin((c-a)/b)])] \cos((x-a)/b) \qquad (7.26)$$

for $a - b\pi/2 \leq x \leq a + b\pi/2$. If truncation occurs at $-c$ in the left-tail, and at $+c$ in the right-tail, the PDF becomes

$$f(x; a, b, c) = \cos((x-a)/b)/\frac{1}{2}[\sin((c-a)/b) - \sin((c+a)/b)]. \qquad (7.27)$$

Using $\sin(C) - \sin(D) = 2\cos((C + D)/2)\sin((C - D)/2)$ this becomes

$$f(x; a, b, c) = \cos((x - a)/b)/[\cos(c/b)\sin(a/b)], \quad \text{for } -c < x < c. \tag{7.28}$$

7.4.1   SIZE-BIASED COSINE DISTRIBUTION

Size-biased cosine distribution (SBCD) can be obtained for any of the forms given above. Consider $f(x; a, b) = 1/(2b)\cos((x - a)/b)$, $a - b\pi/2 \leq x \leq a + b\pi/2$, $b > 0$. Its size-biased versions is $f(x; a, b) = ((cx + d)/(2b))\cos((x - a)/b)$, $a - b\pi/2 \leq x \leq a + b\pi/2$, $b > 0, c, d$ real. Integrate over the range to get

$$\int_{a-b\pi/2}^{a+b\pi/2} f(x)dx = \frac{1}{2b}\int_{a-b\pi/2}^{a+b\pi/2} (cx + d)\cos((x - a)/b)dx = ac + d. \tag{7.29}$$

From this we obtain SBCD as

$$f(x; a, b, c, d) = \frac{(cx + d)}{2b(ac + d)}\cos((x - a)/b), \quad a - b\pi/2 \leq x \leq a + b\pi/2, \ b > 0. \tag{7.30}$$

Higher-order weights can be used to get a variety of other weighted distributions.

7.4.2   TRANSMUTED COSINE DISTRIBUTION

Transmuted distributions were discussed in Chapter 1. This has CDF of the form $G(x; \lambda) = (1 + \lambda)F(x) - \lambda F^2(x)$. From this the transmuted cosine distribution CDF follows as

$$F(x; \lambda) = (\frac{1}{2} + \frac{1}{\pi}\arcsin(x)) + (1 + \lambda\left(\frac{1}{2} - \frac{1}{\pi}\arcsin(x)\right) \tag{7.31}$$

for $-1 < x < 1$ and $\lambda > 0$ is the shape parameter. Differentiate w.r.t. x to get the PDF as

$$f(x; \lambda) = \frac{1}{\pi\sqrt{1 - x^2}}\left(1 - \frac{2\lambda}{\pi}\arcsin(x)\right). \tag{7.32}$$

To show that this is indeed a PDF, integrate over [-1,1] to get

$$\int_{-1}^{+1} f(x)dx = \int_{-1}^{+1} \frac{1}{\pi\sqrt{1 - x^2}}\left(1 - \frac{2\lambda}{\pi}\arcsin(x)\right)dx. \tag{7.33}$$

Put $u = 2\lambda\arcsin(x)$ so that $du = 2\lambda/\sqrt{1 - x^2}$. The limits are from $[\sin(-1/(2\lambda)), \sin(1/(2\lambda))]$. Break the integral into two parts, cancel out common factors and integrate each term to get 1, showing that this is indeed a PDF.

Skewed cosine distribution has PDF $f(x) = 2g(x)G(ax)$, where $g()$ and $G()$ are the PDF and CDF of cosine distribution. Other generalizations include beta-generalized, Kumarasamy, Marshal–Olkin, and McDonalds cosine distributions.

Example 7.1   Mean deviation of cosine distribution   Find the mean deviation of the cosine distribution using the Power method discussed in Chapter 1.

Solution 7.2   We have seen above that the CDF is $\frac{1}{2}[\sin((x-a)/b)+1]$. As the mean is a, we get the MD as

$$MD = 2\int_{ll}^{a} F(x)dx = 2\int_{a-b\pi/2}^{a} \frac{1}{2}[\sin((x-a)/b)+1]\,dx. \tag{7.34}$$

Put $(x-a)/b = t$ so that $dx = b\,dt$. This gives $MD = b\int_{-\pi/2}^{0}(\sin(t)+1)dt = b[\pi/2-1]$.

7.5   SPECIAL COSINE DISTRIBUTIONS

The power cosine distribution has PDF

$$f(\theta;n) = C\cos(\theta)^n\sin(\theta). \tag{7.35}$$

The CDF is

$$F(\theta;n) = 1 - \cos(\theta)^{n+1} \tag{7.36}$$

from which the quantile function follows as $p = F^{-1}(u) = \arccos((1-u)^{(n+1)/2})$. As $1-U$ and U are identically distributed, we could generate random numbers from this distribution as $x = \arccos(u)^{(n+1)/2}$. A squared-cosine distribution with PDF

$$f(\theta;a,b) = \frac{2^{2b-1}\Gamma^2(b+1)}{\pi\Gamma(2b+1)}\cos^{2b}((\theta-a)/2), \tag{7.37}$$

where the parameter b is the angular directional spreading in radar applications in which b is a function of the wave frequency. The four-parameter generalized raised cosine distribution (GRCD) reported by Ahsanullah and Shakil (2019) [2] has PDF

$$f(x;a,b,c,d) = K[c + d\cos(\pi(x-a)/b)],$$
$$a-b \leq x \leq a+b, \; b > 0, 0 < |d| \leq |c| < \infty, \tag{7.38}$$

where $K = 1/(2cd)$ is a constant, $a-b < x < a+b$. This has CDF

$$F(x;a,b) = \frac{1}{2}[1 + (x-a)/b + \frac{d}{c\pi}\sin(\pi(x-a)/b)], \tag{7.39}$$

and the characteristic function is $\phi(t;a,b) = [a\pi^2 - b^2t^2(d-c)]\sin(bt)/[cbt(\pi^2 - b^2t^2)]\exp(iat)$. As a is the location parameter, the mean = median = mode = a. The inflection points are given by $a \pm b/2$. See Ahsanullah, Shakil, and Kibria (2019) [2] for characterizations, percentile points, and parameter estimation. The half-cosine distribution is preferred when error in measurements have symmetric bounding limits but do not tend to aggregate closely at the mid-point (mean). It is less peaked than the cosine distribution, and is midway between a symmetric triangular and a quadratic distributions without the drawback of being discontinuous at the extremes.

7.6   APPLICATIONS

Fast moving molecules bounce off surfaces upon impact either specularly or diffusely. This depends on surface properties like smoothness, temperature, density, conductance, etc., and the angle of incidence. Diffuse reflections are more common than specular reflection. Angular distribution of the molecules can be obtained using fixed directions of a frame of reference. The cosine distribution is used to model angular scattering of molecules in motion around a specularly reflected incident particle direction.[1] Under ideal conditions, the flux of outgoing molecules is proportional to the cosine of the angle from the surface normal. Pressure when applied uniformly in the circumferential (periphery) direction of a canister also results in the cosine law. It is used in the design of compression-only bearing loads to cylinders and rings when cylindrical surfaces have a straight centerline and a constant radius, optimal flow of fluid and vapor-transport systems, and in vacuum operated machines. The Physical Vapor Deposition (PVD) process used in thin-film coating industry is approximately a cosine distribution when the mean free path (MFP) of the evaporant molecules exceeds the scattering depth of the residual atmosphere. Similarly, the angular distribution of sputtered materials from some metal (like copper, zinc) upon bombardment with high-energy ions to an ideally flat surface goes from approximately "below-cosine" to "above-cosine" distribution with increasing ion-energy (more particles are sputtered in the perpendicular direction to the surface, and less in off-normal direction).[2]

It is also used in the study of sputtered atoms transport from nanocomposite coatings to the substrate when near-normal ion bombardment takes place at high pressure, in 2D Laser Detection and Ranging (LADAR), solid-state lasers, and photo-voltaic device designs where reflected and scattered light are assumed to follow an approximate cosine law. It can also be used to model the average monthly temperatures and daylight durations (in hours) of regions in sub-tropics (Ahsanullah, Shakil, and Kibria (2019) [2] fitted a four-parameter GRCD to average monthly temperature of Denver, Colorado; and monthly daylight hours of Nashville, Tennessee).

7.6.1   DIGITAL SIGNAL PROCESSING

DSP uses filters to discard or reduce noise in data. Filtering restricts transmitted bandwidth without significantly affecting quality of signals. The most popular filters for microwave and antenna communications are raised-cosine, square-root raised-cosine and Gaussian filters. Depending on the signal characteristics, the raised-cosine filter can be used as a low-pass filter (the Lanczos-filter is used for high-pass filtering). It belongs to the class of Nyquist filters, and is used extensively for pulse shaping in modems designed for both wired and wireless (digital) systems. As its magnitude response is of the cosine wave form, it is also called a comb filter (for $M = 2$). The time response of this filter goes through zero with a period that exactly matches the symbol rate. The impulse response of a raised cosine filter is $h(t) = \frac{\sin(\pi t/T)}{\pi t/T} \frac{\cos(b\pi t/T)}{1-(2bt/T)^2}$. The

---

[1]Maxwell energy and cosine angular distributions are used in most engineering fields.
[2]It is called "cosine emission" in some fields.

corresponding frequency function resembles the cosine law:

$$H(f) = \frac{T}{2}\left[1 + \cos(\frac{\pi T}{b}(|f| - \frac{1-b}{2T}))\right] \quad \text{for} \quad \frac{1-b}{2T} < |f| < \frac{1+b}{2T}, \tag{7.40}$$

where T is the symbol period. It reduces the Inter-Symbol Interference (ISI) and also smoothen sharp edges of digital baseband signals (called pulse-shaping filter). This step is usually performed before modulation of sampled signals so as to reduce out-of-band frequency content. The far-out side-lobes of the cosine filter decay inverse quadratically as $1/t^2$, so that a cosine-squared distribution with decay rate $1/t^3$ is used in some applications. The half-symmetry model (called square-root raised-cosine filter) splits the filter with one half in the transmit path (Tx), and the other half in the receiver path (Rx), so that the combined response is that of a Nyquist filter. In other words, the pulse shape observed after the signals pass through the Rx filter is cosine distributed and satisfies the Nyquist criterion.

7.6.2   PATH PLANNING IN ROBOTICS

Moving robots, self-driving autonomous vehicles, and unmanned aerial vehicles (UAV) use a compass, multiple cameras, and depth-perception sensors to plan collision-free credible path forward. Without loss of generality, we will confine ourselves to robotic motion. If the current state of a robot moving in flat 2D terrain is likely to result in a collision with an object, the present position is contained in a neighborhood circle of "obstacle points" with its center at the centroid of the robot, and radius being a function of the obstacle's size, shape, distance, and dynamics of motion. The circular region becomes elliptical when the obstacle is in deterministic motion with its major axis perpendicular to the line connecting the centroids (LCC) of the robot and the obstacle (major axis will coincide with the LCC and minor axis will shrink when the obstacle is moving away along a line parallel to the LCC, and vice versa).

The shape is spherical (for static obstacles) or ellipsoidal (for deterministic moving obstacles) in 3D motion (like self-propelled drones, UAV and in robotic surgeries, image-guided medical needle steering, targeted on-site treatments like drug delivery, radiation dose at tumor sites, etc.). The encircling region will be called the "boundary region" (BR) irrespective of its shape. Alternate collision-free paths may have to be worked out using this information, which could result in a set of equivalent paths. Robotics use "locality models" to find plausible alternate paths with respective probabilities assigned to them using the set of all known collisions. Probability of collision will drastically decrease as per the raised cosine law $\frac{1}{2b}[1 + \cos(\pi r)/2b]$, for $0 \le r \le 2b$ (in 2D) when the robot is radially moved from current location (where probability of collision peaks) toward some points on both sides of the perpendicular to the LCC on the BR (bivariate extensions are used for 3D motion where path diversity is more). The challenge is to locate few such paths using conditional probability of outcomes of a set of collisions. A half-region (like circumference of a semi-circle) is identified toward the obstacle on the BR and a few random points on the boundary generated (we could ignore the other half-boundary when the robot motion is forward, and use the arcsine law to generate the random points so that less number of points are generated in the proximity of intersection point of the BR and LCC). The half-region reduces to a smaller sector when the obstacle is moving away in more-or-less the same direction, and a larger sector than it otherwise (collision chances increase when the obstacle moves toward the robot so that the BR will continuously grow in size for a safe escape path).

Linear (first-order) locality models abstract away size and shape information, and build a rapidly computable or continuously updatable incremental path resulting in a bounded path-sequence. The risk (probability) of collision is then computed for each such point, and sorted in ascending order. If the arcsine law is used for random point generation, there will be less number of points on paths most closely approaching the obstacle with high risks. Each of these points are tagged with the current centroid (robot position) and saved (say stacked so that minimum risk is at the top (which can be efficiently implemented using a min-heap)). If the direction of robotic motion is unrestricted (as in robotic-boats and rescue missions on water surfaces or robots in playfields) and the obstacle is static, we could approximate the bounded radial path by a straight line. Then the probability of collision (in 2D) is the ratio of the area of a sector of the BR formed by two neighboring points to the total area of the BR (e.g., area of semi-circle). The saved points may be used when backtracking is necessary. This is especially true when an equivalent class of safe paths exist. The entire process is repeated after the robot is moved to the most plausible position, resulting in a collision-free trajectory.

7.6.3   ANTENNA DESIGN

Antennas are integral parts of wireless communication systems. They are the part-and-parcel of cell phones, WiFi and Bluetooth devices, smart-meters, RFID tags used in automobiles and inventory systems, GPS devices, radars, command and control systems, deep space communication systems, and in robotics. They convert signals generated by a circuit into wave form that gets transmitted over a channel to be received by another antenna that does the reverse job of converting received electromagnetic signals back into original form. Operations like encryption and compression may happen during this transmission in secure communication systems. The size of the antenna depends on the frequency of operation. Mathematically, the antenna size is inversely proportional to the frequency of signals (fairly large at lower frequencies as in deep-space communication and smaller at higher frequencies as in cell phones). The continuous uniform and cosine distributions are the most popular statistical distributions used in antenna design for synthesis and tolerance analysis of received signals. It is used to approximate the beam width of aperture radars,

$$f(x; a, PD) = PD + (1 - PD)\cos^2(\pi x/a), \quad -a/2 < x < a/2, \tag{7.41}$$

where PD is the voltage pedestal level and a is a constant.

7.7 SUMMARY

This chapter introduced the cosine distribution and its variants. This distribution is a favorite choice in communications engineering, antenna design, and robotics due to the peculiar shape of the PDF. Important properties of this distribution are derived and extensions of it briefly discussed. These are extensively used in industrial experiments that involve the flow of compressed fluids and gases, in beam-scattering applications of surface-chemistry and photo-voltaic experiments, for optimal performance of vacuum operated machinery, and in the design of optimal bearing loads of cylinders and rings. A brief discussion of truncated, size-biased, and transmuted cosine distributions also appears. The chapter ends with several useful practical applications of this distribution in electronics, communication engineering, etc.

CHAPTER 8

Normal Distribution

8.1 INTRODUCTION

The normal distribution is perhaps the most widely studied continuous distribution in statistics. It originated in the works of Abraham De Moivre (1733), who derived it as the limiting form of binomial distribution BINO(n,p) as $n \to \infty$ where p remains a constant. It is known by the name Gaussian distribution in engineering in honor of the German mathematician Carl Friedrich Gauss (1777–1855) who applied it to least square estimation technique, and worked extensively with its properties.[1] It has two parameters, which are by convention denoted as μ and σ to indicate that they capture the location (mean) and scale information.

The PDF is denoted by $\phi(x; \mu, \sigma)$ (and the CDF by $\Phi(x; \mu, \sigma)$) as

$$\phi(x; \mu, \sigma) = \frac{1}{\sigma\sqrt{2\pi}} e^{-\frac{1}{2}(\frac{x-\mu}{\sigma})^2}, \quad -\infty < x < \infty, \ -\infty < \mu < \infty, \ \sigma > 0. \quad (8.1)$$

The distribution is denoted by $N(\mu, \sigma^2)$ where the first parameter is always the population mean, and the second parameter is the population variance. The first parameter should be specified even when the mean is zero. Thus, $N(0, \sigma^2)$ denotes a normal distribution with zero mean. A Gaussian distributed random variable is called a normal deviate. Some authors use the notation $N(\mu, \sigma)$ where the second parameter is the population standard deviation, and vertical bar (|) instead of semicolon as $\phi(x|\mu, \sigma)$. It is called a *normalized normal, standard normal, standardized normal,* or *unit normal* distribution (UND) when the mean is 0 and variance is 1. (Note that the abbreviation SND is used for skew-normal distribution discussed in Section 8.5.) The standard normal random variable will be denoted by $Z(0, 1)$ or Z; its PDF by $\phi(z; 0, 1)$, $\phi(z)$, or $f(z)$; and CDF by $\Phi(z; 0, 1)$ or $\Phi(z)$ where z in the CDF is the upper limit of the integral.

Any normal distribution (with arbitrary μ and σ) can be converted into the standard normal form $N(0,1)$ using the transformation $Z = (X - \mu)/\sigma$. This is called "standard normalization" as it standardizes the variate X to have zero mean and variance unity. The reverse transformation is $X = Z\sigma + \mu$. This shows that the tail areas of an arbitrary normal distribution can be obtained from the table of standard normal.

Example 8.1 Probability of normal deviates The radius of a batch of pipes is known to be normally distributed with mean .5 inch and variance .009. What proportions of a batch of 132 pipes have radius more than 2 standard deviations in the higher side?.

---

[1]This distribution was known by various other names like "law of error," "second law of Laplace," etc., before it came to be called Gaussian distribution.

Solution 8.2   As radius $\sim N(.5, 0.009)$, standard deviation is 0.0948683. Standard normalize it to get $Z = (X - 0.5)/0.0948683$. Area above two standard deviations for $N(0,1)$ is $1 - 0.9772 = 0.0228$. Thus, in a batch of 132 pipes we expect $132*0.0228 = \lfloor 3.0096 \rfloor = 3$ pipes to have radius more than 2 standard deviations.

8.1.1   ALTERNATE REPRESENTATIONS

The PDF can also be written as

$$f(x; \mu, \sigma) = \frac{1}{\sqrt{2\pi\sigma^2}} \exp(-(x - \mu)^2/(2\sigma^2)), \quad |\mu| < \infty, \ \sigma > 0, \ I_{-\infty,\infty}(z), \quad (8.2)$$

where $I_{-\infty,\infty}(z)$ is the indicator function. As $\sqrt{2\pi} = 2.506628$, it can also be written as

$$\begin{aligned} \phi(x; \mu, \sigma) &= 1/(2.506628\sigma) \exp(-(1/2)((x - \mu)/\sigma)^2) \\ &= (0.39894/\sigma) \exp(-0.5((x - \mu)/\sigma)^2), \quad |x| < \infty, \end{aligned} \quad (8.3)$$

and the UND is denoted as

$$\phi(z) = (1/\sqrt{2\pi}) \exp(-z^2/2) = 0.39894 \ \exp(-z^2/2). \quad (8.4)$$

The transformation $Y = Z/\sqrt{2}$ results in the PDF $f(y) = \exp(-y^2)/\sqrt{\pi}$, which was used by Gauss in his works. Some engineering fields use the substitution $h = 1/(\sqrt{2}\sigma)$ (called precision modulus) to get the PDF in the form

$$\phi(x; \mu, h) = (h/\sqrt{\pi}) \exp(-h^2(x - \mu)^2) = (0.56418958 \ h) \exp(-[h(x - \mu)]^2), |x| < \infty. \quad (8.5)$$

8.2   RELATION TO OTHER DISTRIBUTIONS

If X and Y are IID normal variates with zero means, then $U = XY/\sqrt{X^2 + Y^2}$ is normally distributed ([134], [111]). In addition, if $\sigma_x^2 = \sigma_y^2$, then $(X^2 - Y^2)/(X^2 + Y^2)$ is also normally distributed. Product of independent normal variates has a Bessel-type III distribution. The square of a normal variate is gamma distributed (of which χ^2 is a special case). In general, if $X_1, X_2, \ldots, X_k$ are IID $N(0, \sigma^2)$, then $\sum_{i=1}^{k} X_i^2/\sigma^2$ is χ_k^2 distributed. If $Z_1 = X_1^2/(X_1^2 + X_2^2)$, $Z_2 = (X_1^2 + X_2^2)/(X_1^2 + X_2^2 + X_3^2)$, and so on, $Z_j = \sum_{i=1}^{j} X_i^2 / \sum_{i=1}^{j+1} X_i^2$, then each of them are Beta-I distributed, as also the product of any consecutive set of Z_js are beta distributed [75].

If X and Y are independent normal random variables then X/Y is Cauchy distributed (Chapter 9). If X is chi-distributed (i.e., $\sqrt{\chi_m^2}$) and Y is independently distributed as $Beta((b - 1)/2, (b - 1)/2)$, then the product $(2Y - 1) X$ is distributed as $N(0,1)$ [57].

There are many other distributions which tend to the normal distribution under appropriate limits. For example, the binomial distribution tends to the normal law when the sample size

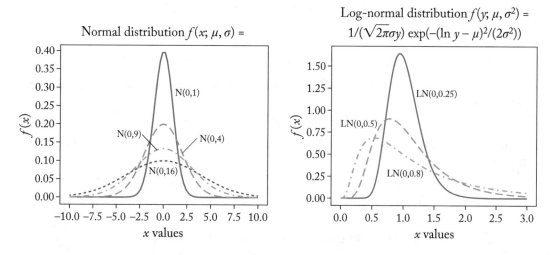

Figure 8.1: Normal and lognormal distributions.

n becomes large. The convergence is more rapid when $p \to 1/2$ and n is large. The lognormal and normal distributions are related as follows: If $Y = \log(X)$ is normally distributed, then X has lognormal distribution. The lognormal distribution can be obtained from a normal law using the transformation $X = \exp(Y)$ (page 108). Normal distribution is also related to Student's t, Snedecor's F, and Fisher's Z distributions as shown in later chapters. See Roy (2003) [119] for a discrete normal distribution.

8.3   PROPERTIES OF NORMAL DISTRIBUTION

The general normal distribution has two parameters, the second of which is positive (the first parameter by convention is the mean μ). Variance of the population (usually indicated by σ^2) controls the shape and significant width from the mean. Although the distribution extends from $-\infty$ to $+\infty$, smaller values of σ^2 results in narrower width and larger values results in wider widths (Figure 8.1). In the limit when $\sigma^2 \to 0$, this converges to the Dirac's delta function. The distribution is symmetric about the mean with relatively shorter tails than Cauchy and Student's t distributions. Due to symmetry $\Phi(-c) = 1 - \Phi(c)$ for $c > 0$, and $\Phi(0) = 1/2$, so that median=mean=mode with modal value $M = 1/[\sigma\sqrt{2\pi}]$. In terms of the modal value, the PDF takes the simple form $\phi(z) = M \exp(-z^2/(2\sigma^2))$. If $c < d$, the area from c to d can be expressed as $\Phi(d) - \Phi(c)$. Any linear combination of IID normal variates is normally distributed. Symbolically, if $X_1, X_2, \ldots, X_k$ are IID $N(\mu_i, \sigma_i^2)$, then $Y = \mp c_1 X_1 \mp c_2 X_2 \mp \ldots \mp c_k X_k = \sum_{i=1}^{k} \mp c_i X_i$ is normally distributed with mean $\sum_{i=1}^{k} \mp c_i \mu_i$ and variance $\sum_{i=1}^{k} c_i^2 \sigma_i^2$. This is most easily proved using the MGF or ChF. An immediate consequence of this result is that any normal distribution $N(\mu, \sigma^2)$ can be decomposed into k independent components $X_1, X_2, \ldots, X_k$ where $X_j \sim N(\mu/k, \sigma^2/k)$.

8.3.1   MOMENTS AND GENERATING FUNCTIONS

As the distribution is symmetric, all odd central moments except the first are zeros. The even moments are given by

$$\mu_{2k} = (2k!)(\sigma^2)^k / [2^k k!]. \tag{8.6}$$

This can easily be proved using gamma integrals. The mean deviation is $\sigma \sqrt{2/\pi}$. The MGF is easily obtained as

$$M_x(t) = E(e^{tx}) = \int_{-\infty}^{\infty} \frac{1}{\sigma \sqrt{2\pi}} e^{tx} e^{-\frac{1}{2}((x-\mu)/\sigma)^2} dx. \tag{8.7}$$

Put $z = (x - \mu)/\sigma$ in the above so that $dz = dx/\sigma$ and $x = \mu + \sigma z$, to get

$$M_x(t) = (1/\sqrt{2\pi}) \int e^{t(\mu + \sigma z)} e^{-z^2/2} dz = e^{t\mu}/\sqrt{2\pi} \int e^{t\sigma z - \frac{1}{2} z^2} dz. \tag{8.8}$$

Write the exponent as $-\frac{1}{2}(z - t\sigma)^2 + \frac{1}{2} t^2 \sigma^2$. As $e^{\frac{1}{2} t^2 \sigma^2}$ is constant, take it outside the integral to get

$$e^{t\mu + \frac{1}{2} t^2 \sigma^2} (1/\sqrt{2\pi}) \int_{-\infty}^{\infty} e^{-(z - t\sigma)^2/2} dz. \tag{8.9}$$

As the integral evaluates to $\sqrt{2\pi}$, we get the desired result $M_x(t) = e^{t\mu + \frac{1}{2} t^2 \sigma^2}$. The quantiles are given by $Q_p = \mu + \sqrt{2}\sigma \, \mathrm{erf}^{-1}(2p - 1)$. The points of inflection are $\mu \pm \sigma$. See Table 8.1 for further properties.

The normal distribution is extensively used in statistical inference. It is the basis of many procedures that include confidence intervals for unknown parameters, prediction intervals for future observations, tests of various hypotheses, and estimation of parameters. It is applicable for approximation due to central limit theorem, even in non-normal situations when the sample size is quite large. This fact is used in many applied fields like fluid dynamics and crystallography to approximate the distance of electron movement in a disordered system with impurities present. Given below is a theorem about the mean of a random sample from a normal population, which is used in CI construction and tests about the mean.

Theorem 8.3   *The sample mean $\bar{x}_n$ of any sample of size $n \geq 2$ from a normal population $N(\mu, \sigma^2)$ is itself normally distributed as $N(\mu, \sigma^2/n)$.*

Proof. The easiest way to prove this result is using the MGF (or ChF) as follows. Let $M_x(t)$ be the MGF of a normal distribution. Then $M_{\bar{x}}(t) = [M_x(t/n)]^n = [e^{\mu t/n + \frac{1}{2}(t/n)^2 \sigma^2}]^n = e^{n\mu t/n + \frac{1}{2} n\sigma^2 (t/n)^2} = e^{\mu t + \frac{1}{2} t^2 (\sigma^2/n)}$, which is the MGF of a normal distribution with mean μ and variance σ^2/n. This proves the result.

Table 8.1: Properties of normal distribution ($\frac{1}{\sigma\sqrt{2\pi}}e^{-\frac{1}{2}(\frac{x-\mu}{\sigma})^2}$)

Property	Expression	Comments
Range of X	$-\infty < x < \infty$	Infinite
Mean	μ	
Median	μ	Mode = μ
Variance	σ^2	
Skewness	$\gamma_1 = 0$	Symmetric
Kurtosis	$\beta_2 = 3$	for $N(0, 1)$
MD	$E\lvert X - \mu\rvert = \sigma\sqrt{2/\pi}$	$2\int_{-\infty}^{\mu}\Phi(x)dx$
CDF	$\Phi((x - \mu)/\sigma)$	
Absolute	$E(\lvert X\rvert^r) = (r-1)!!\sigma^r$	for $r = 2k$
Moments	$E(\lvert X\rvert^r) = \sqrt{2/\pi}2^k k!\sigma^r$	for $r = 2k + 1$
Moments	$\mu_{2r} = \sigma^{2r}(2r)!/[2^r r!]$	Even
MGF	$\exp(t\mu + \frac{1}{2}t^2\sigma^2)$	
ChF	$\exp(it\mu - \frac{1}{2}t^2\sigma^2)$	
Additivity	$N(\mu_1, \sigma_1^2) \pm N(\mu_2, \sigma_2^2)$, IID	$N(\mu_1 \pm \mu_2, \sigma_1^2 \pm \sigma_2^2)$

$\mathrm{erf}(z) = \frac{1}{\sqrt{\pi}}\int_0^z e^{-x^2/2}dx = \frac{2}{\sqrt{\pi}}\int_0^z e^{-t^2}dt$

8.3.2 TRANSFORMATIONS TO NORMALITY

Normality of parent population is a fundamental assumption in many statistical procedures. For example error terms in regression are assumed to be normally distributed with zero mean. Normality tests of the sample mean can reveal whether the data came from a normal distribution or not. When the data are not normally distributed, a simple transformation may sometimes transform it to nearly normal form. For instance, count data are usually transformed using the square root transformation, and proportions are transformed using logit transformation as $y = \frac{1}{2}\log(p/q)$ where $q = 1 - p$.

Example 8.4 Probability $\Pr(X \leq Y)$ **for two populations** $N(\mu, \sigma_i^2)$ If X and Y are independently distributed as $N(\mu, \sigma_i^2)$, for $i = 1, 2$ find the probability $\Pr[X \leq Y]$.

Solution 8.5 As X and Y are IID, the distribution of $X - Y$ is $N(0, \sigma_1^2 + \sigma_2^2)$. Hence, $\Pr[X \leq Y] = \Pr[X - Y \leq 0] = \Phi(0) = 0.5$, irrespective of the variances.

8.3.3   TAIL AREAS

The CDF of UND is given by

$$F_z(z_0) = \Phi(z_0) = \frac{1}{\sqrt{2\pi}} \int_{-\infty}^{z_0} e^{-z^2/2} dz. \tag{8.10}$$

If $X \sim N(\mu, \sigma^2)$ then $\Phi(z) = F((x - \mu)/\sigma)$. Due to symmetry around the mean, $\Phi(-z) = 1 - \Phi(z)$, or equivalently $F(\mu - x) = 1 - F(\mu + x)$. This can be approximated as $\Phi(z) = 0.5 + (1/\sqrt{2\pi})(z - z^3/6 + z^5/40 - \cdots = 0.5 + (1/\sqrt{2\pi}) \sum_{k=0}^{\infty}(-1)^k z^{2k+1}/[(2k+1)k!2^k]$. Because of the symmetry of the normal curve, the CDF is usually tabulated from 0 to some specified value x. The error function erf() is defined as

$$\text{erf(z)} = \frac{1}{\sqrt{\pi}} \int_0^z e^{-x^2/2} dx = \frac{2}{\sqrt{\pi}} \int_0^z e^{-t^2} dt. \tag{8.11}$$

Using a simple change of variable, this can be expressed in terms of the incomplete gamma integral as $\sqrt{\pi}\gamma(1/2, z^2)$. The complement of the above integral is denoted by $\text{erfc}(z)$. The error function $\text{erfc}(z)$ has an infinite series expansion as

$$\text{erf(z)} = \frac{2}{\sqrt{\pi}} \int_0^z e^{-t^2} dt = \frac{2}{\sqrt{\pi}} e^{-z^2} \sum_{k=0}^{\infty} 2^k z^{2k+1}/[1.3.\ldots(2k+1)]. \tag{8.12}$$

This series is rapidly convergent for small z values. The $\text{erfc}(z)$ can be expressed in terms of confluent hypergeometric functions as

$$\text{erf(z)} = \frac{2z}{\sqrt{\pi}} {}_1F_1(1/2, 3/2, -z^2) = \frac{2z}{\sqrt{\pi}} e^{-z^2} {}_1F_1(1, 3/2, z^2). \tag{8.13}$$

When the ordinate is in the extreme tails, another approximation as

$$\Phi(-x) = \frac{\phi(x)}{x} \left(1 - 1/x^2 + 3/x^4 - \cdots + (-1)^n 3 * 5 * \cdots (2n - 1)/x^{2n}\right) \tag{8.14}$$

can be used. Replace $-x$ by $+x$ to get an analogous expression for right tail areas. Scaled functions of the form $C \exp(-dx^2)$ are quite accurate to approximate the tail areas.

Problem 8.6   If X and Y are normally distributed with zero means, correlation ρ and variances σ_1^2 and σ_2^2, show that the distribution of $(X/\sigma_1 + Y/\sigma_2)$ and $(X/\sigma_1 - Y/\sigma_2)$ are independent normal variates.

Example 8.7 Mean deviation of Normal distribution   Find the mean deviation of the Normal distribution.

Table 8.2: Area under normal variates

Sigma Level	Area As %	Outside Range (ppm)	Sigma Level	Area As %	Outside Range (ppm)
$\mp 1\sigma$	68.26	317,400	$\mp 1.5\sigma$	86.64	133,600
$\mp 2\sigma$	95.44	45,600	$\mp 2.5\sigma$	98.76	12,400
$\mp 3\sigma$	99.74	2600	$\mp 3.5\sigma$	99.96	400
$\mp 4\sigma$	99.99366	63.40	$\mp 5\sigma$	99.9999426	0.574
$\mp 6\sigma$	99.9999998	0.002	$> 6\sigma$	99.99999999	Negligible

Solution 8.8 Let $X \sim N(\mu, \sigma^2)$. As the normal distribution tails off to zero at the lower and upper limits, the MD is given by

$$MD = 2 \int_{ll}^{\mu} F(x)dx = 2 \int_{-\infty}^{\mu} \Phi(x)dx. \tag{8.15}$$

Put $z = (x - \mu)/\sigma$, so that $dx = \sigma\, dz$. The lower limit remains the same, but the upper limit becomes 0. Thus, we get

$$MD = 2\sigma \int_{-\infty}^{0} \Phi(z)dz. \tag{8.16}$$

This integral can readily be evaluated using integration by parts. Put $u = \Phi(z)$ and $dv = dz$, so that $du = \phi(z)$. This gives

$$MD = 2\sigma \left[z\Phi(z) \mid_{-\infty}^{0} - \int_{-\infty}^{0} z\phi(z)dz \right] = 2\sigma \int_{0}^{\infty} z\phi(z)dz. \tag{8.17}$$

Substitute for $\phi(z) = (1/\sqrt{2\pi}) \exp(-z^2/2)$ and use $\int_{0}^{\infty} z^{2n+1} \exp(-z^2/2)dz = n!2^n$. Then the integral becomes $1/\sqrt{2\pi}$. Apply the constant 2σ to get the

$$MD = 2\sigma * 1/\sqrt{2\pi} = \sigma\sqrt{2/\pi}. \tag{8.18}$$

Problem 8.9 Prove that the sum of squares of two IID standard normal variates is exponentially distributed.

Solution 8.10 Let X and Y be distributed as $N(0,1)$ and $Z = X^2 + Y^2$. The CDF of Z is $\Pr[Z \le z] = \Pr[X^2 + Y^2 \le z]$. As $X^2 + Y^2 = r$ is the equation of a circle, the above probability is the probability content of bivariate normal distribution within a circle of radius $\sqrt{r}$. This is given by $G(y) = \int \int_{x^2+y^2 \le r} \frac{1}{2\pi} \exp(-(x^2 + y^2)/2)dxdy$. Put $x = r\cos(\theta)$ and $y = r\sin(\theta)$, so that $r = \sqrt{x^2 + y^2}$. Then

$$G(y) = \int_{\theta=0}^{2\pi} \int_{r=0}^{\sqrt{y}} \frac{1}{2\pi} \exp(-r^2/2) r \, dr \, d\theta. \tag{8.19}$$

Integrate w.r.t. θ, and cancel out 2π to get $G(y) = \int_{r=0}^{\sqrt{y}} \exp(-r^2/2) r \, dr = 1 - \exp(-y/2)$. This is the CDF of a standard exponential distribution (or special case of χ^2 distribution with 2 DoF).

8.3.4   ADDITIVITY PROPERTY

As mentioned in Section 8.3, linear combinations of IID normal variates are normally distributed. If $X \sim N(\mu_1, \sigma_1^2)$ and $Y \sim N(\mu_2, \sigma_2^2)$ are independent, then $X \pm Y \sim N(\mu_1 \pm \mu_2, \sigma_1^2 + \sigma_2^2)$. This is an important point in practical experiments, where results from two or more processes that are independent and normally distributed need to be combined.

Example 8.11   Linear combination of IID normal variates $X \sim N(10,3)$, $Y \sim N(15,6)$, and $Z \sim N(9,2.5)$, find the mean and variance of the following functions: (i) $U = X - 2Y + 3Z$ and (ii) $V = 2X - 1.2Y - Z$.

Solution 8.12   Use the linear combination property to get $E(U) = 10 - 2 * 15 + 3 * 9 = 37 - 30 = 7$, $Var(U) = 3 + 4 * 6 + 9 * 2.5 = 49.5$ so that $U \sim N(7, 49.5)$. In the second case, $E(V) = 2 * 10 - 1.2 * 15 - 9 = -7$ and $Var(V) = 4 * 3 + 1.44 * 6 + 2.5 = 23.14$.

8.4   SPECIAL NORMAL DISTRIBUTIONS

There are several special cases of this distribution that includes truncated normal, lognormal, skewed normal, and inverse normal.

8.4.1   TRUNCATED NORMAL DISTRIBUTION

The PDF of an asymmetrically truncated normal distribution with truncation points a and b is

$$f(x; \mu, \sigma, a, b) = \frac{1}{\sigma} \phi((x - \mu)/\sigma) / [\Phi((b - \mu)/\sigma) - \Phi((a - \mu)/\sigma)] \text{ for } a < x < b.$$

As the truncation point is asymmetric, the area enclosed is $\int_a^b \phi((x - \mu)/\sigma) dx$. Put $z = (x - \mu)/\sigma$, so that $dz = dx/\sigma$. The limits are changed as $(a - \mu)/\sigma$ and $(b - \mu)/\sigma$. Thus,

$$\int_a^b \phi((x - \mu)/\sigma) dx = \sigma \int_{(a-\mu)/\sigma}^{(b-\mu)/\sigma} \phi(z) dz = \sigma \left[\Phi((b - \mu)/\sigma) - \Phi((a - \mu)/\sigma) \right]. \tag{8.20}$$

Dividing by this quantity gives the PDF as desired. The mean is given by $\mu_t = \mu + \sigma (\phi(h) - \phi(k)) / (\Phi(k) - \Phi(h))$ where $h = (a - \mu)/\sigma$ and $k = (b - \mu)/\sigma$. Although the normal distribution belongs to the LaS family, the above expression shows that truncated versions do not belong to this family. This results in the half-normal distribution (HND) when

$a = \mu = 0$ and $b = \infty$. The HND-shaped curves are used to analyze the spectrum associated with collision-induced absorption of gaseous mixtures (like rare gases) in statistical mechanics as $I(x; \tau) = \mu_0^2 \tau / (2\sqrt{\pi}) \quad \exp(-x^2/(4/\tau^2))$ for $x > 0$, so that $2I(x; \tau)/\mu_0^2$ is half-Gaussian distributed with variance $2/\tau^2$ where τ is the duration of the collision.

The symmetric power distribution has PDF

$$f(x; \mu, \sigma, b) = C \exp(-\frac{1}{2}|(x - \mu)/\sigma|^{2/(1+b)}, \tag{8.21}$$

where $C = \frac{1}{\Gamma(a)2^a \sigma}$ with $a = 1 + (1 + b)/2$. This reduces to the normal law when $b = 0$, so that $a = 3/2$ and $C = \frac{1}{\Gamma(3/2)2^{3/2}\sigma} = 1/[\frac{1}{2}\sqrt{\pi}2^{3/2}]\sigma = 1/[\sigma\sqrt{2\pi}]$. This distribution is leptokurtic when $b > 0$ and platykurtic when $b < 0$, approaches the asymptotic uniform distribution as $b \to -1$.

Problem 8.13   If $X \sim N(0, \sigma)$ find $E(-cX^2)$.

8.5   SKEW-NORMAL DISTRIBUTION

Normality of parent data is a basic assumption in various statistical procedures. Actual data in most applications inevitably exhibit some degree of departure from normality. This is called skewness in statistics. A distribution that is skewed is also called *askew* (which literally means misaligned or uneven). The skewness in sample data indicate systematic deviations from the Gaussian (bell-shaped) law, which is quantified using a skewness measure. Several variables in engineering fields like civil, thermal, metallurgical, and aerospace engineering as well as many economic and financial variables are askew. This may be either to the left of the mean (negatively skewed) or to the right (positively skewed). Examples from actuarial sciences include insurance risks, property liability insurance claims, losses, etc. It also occurs in reliability, survival analysis, Bayesian estimation, and regression analysis. One method to model such skewed data is using "nearly symmetric" distributions. Thus, a χ^2 distribution with very low DoF (like DoF = 3 or 4) may be used to model positively skewed data. Many other skew-symmetric distributions are available in the literature. One promising family is the skew-normal distribution (SND), which is "nearly normal" for small values of the skewness parameter described below. Whereas some parametric statistical distributions approach normality asymptotically as some parameters tend to specific values, the SND includes the normal law as a "proper member." These distributions share some desirable properties of the normal law and simultaneously exhibit slight variations from normality using an additional shape parameter (which is denoted by λ, a, etc.).

The SND indirectly appeared in the works of Roberts (1966) [118], O'Hagan and Leonard (1976) [105], and Aigner and Lovell (1977) [3]. Azzalini (1985, 1986) [14, 15] systematically studied the properties of SND, which aroused widespread interest among both Bayesian and frequentist statisticians, and gave a momentum to the adaptation of skew-normal models among researchers in various disciplines. This allows one to build fairly flexible models that have continuous variation from normality to non-normality and vice versa using an additional parameter. Other terminology for the SN family includes "generalized chi-distribution" (because the square of SND is distributed as χ_1^2 as shown below), and "modulating symmetry" which was advocated by Azzalini recently. To understand the SND, let $\phi(x)$ and $\Phi(x)$ denote the standard normal PDF and CDF, respectively. As the standard normal PDF is symmetric, it trivially follows that $\phi(-x) = \phi(x)$ and $\Phi(x) = 1 - \Phi(-x)$. Now scale the PDF (at x) of a normal distribution by the CDF $\Phi(ax)$ (up to ax) of the same or related distribution (a scaled value of the ordinate x by a suitable constant a) and find the normalizing constant to make the total area into unity. Instead of scaling $\phi(x)$ by $\Phi(ax)$ we could also scale by $S(ax)$ where $S()$ is the survival function, or use the area from 0 to ax to get similar distributions. Define a new PDF as

$$f(x; a) = 2\phi(x)\Phi(ax) = \sqrt{2/\pi} \ \ e^{-x^2/2} \left(1 + \operatorname{erf}(ax/\sqrt{2})\right), \quad \text{for } a, |x| < \infty. \quad (8.22)$$

By expanding $\phi(x)$ and $\Phi(ax)$, this could also be written as $f(x; a) = \frac{1}{\pi} \exp(-x^2/2) \int_{-\infty}^{ax} \exp(-y^2/2) dy$. It is called the one-parameter SND. An interpretation of this result is that the PDF $\phi(x)$ when multiplied by a skew factor $2\Phi(ax)$ gives rise to a new PDF called the SND. Here a is the shape parameter (sometimes called skewness parameter) that regulates the asymmetry, and $2 \Phi(ax)$ is called the "skewness factor." A negatively skewed distribution results when $a < 0$, and a positively skewed distribution for $a > 0$. In other words, the skewness is decided by the absolute value $|a|$, but is usually limited in the interval $(-1, +1)$. As $\Phi(0) = 1/2$, it becomes the UND $Z(0,1)$ for $a = 0$. It approaches a half-normal distribution as $a \to \pm\infty$. Symbolically, as $a \to \infty$, the PDF becomes $f(x) = 2\phi(x)$, $x \geq 0$, and as $a \to -\infty$ the PDF becomes $f(x) = 2\phi(x)$, $x \leq 0$. This is why a is usually kept in the interval $(-1, +1)$. Replace $\Phi(ax) = \frac{1}{2}\left(1 + \operatorname{erf}(ax/\sqrt{2})\right)$ to get the last expression in (8.22).

To prove that it is indeed a PDF, consider the general case (Azzalini (1985)). If $f(x)$ is the PDF of a continuous random variable X symmetric around 0, and $G()$ is an absolutely continuous CDF of a continuous random variable Y that also is symmetric around 0, then $g(x) = 2f(x)G(ax)$ is a valid PDF for any real value of a.

As X and Y are both symmetric around zero, $X - aY$ has zero mean, and is also symmetric around zero so that $\Pr[X - aY < 0 = 1/2]$ because the mean and median coincide for symmetric distributions. Now consider $\Pr(X - aY < 0)$. Using conditional expectation, this can be written as $\Pr(X - aY < 0) = E_y[\Pr[X < ay|Y = y]] = E_y[G(ay)|Y = y]$. As X and Y are continuous, this in integral form is $\int_{-\infty}^{\infty} G(ay) f(y) dy$. Thus, it follows that $2 \int_{-\infty}^{\infty} F(ay) f(y) dy = 1$.

There is another simple way (without conditional expectation) to prove that $2\phi(x)\Phi(ax)$ is indeed a PDF. An astute reader will notice that $\int_{-\infty}^{\infty} \Phi(ax) \phi(x) dx$ is the expected value of $\Phi(ax)$.

Theorem 8.14   *If Z is an N(0,1) random variable and a, b are any real constants, then $E[\Phi(aZ + b)] = \Phi(b/\sqrt{1 + a^2})$ where $\Phi()$ denotes the CDF of SND.*

Proof. By definition $E[\Phi(aZ + b)] = \int_{-\infty}^{\infty} \Phi(az + b)\phi(z)dz$. Differentiate w.r.t. the free variable b (so that $\phi(z)$ is a constant) to get $\frac{\partial}{\partial b} E[\Phi(aZ + b)] = \int_{-\infty}^{\infty} \phi(az + b)\phi(z)dz$. Now substitute the PDF inside the integral to get the RHS as $\frac{1}{2\pi} \int_{-\infty}^{\infty} \exp\left[-\frac{1}{2}((az + b)^2 + z^2)\right] dz$. Write $a^2z^2 + z^2$ as $z^2(1 + a^2)$, divide by $(1 + a^2)$, and complete the square to get $\frac{1}{2\pi} \exp(-b^2/[2(1 + a^2)]) \int_{-\infty}^{\infty} \exp(-\frac{1+a^2}{2}(z + ab/(1 + a^2))^2)dz$. Put $x = \sqrt{1 + a^2}(z + ab/(1 + a^2))$ so that $dx = \sqrt{1 + a^2}dz$ to get the RHS as $\frac{1}{\sqrt{2\pi}\sqrt{1+a^2}} \exp(-b^2/[2(1 + a^2)])$. This in terms of $\phi()$ notation is $\frac{\partial}{\partial b} E[\Phi(aZ + b)] = \frac{1}{\sqrt{1+a^2}}\phi(b/\sqrt{1 + a^2})$. Integrate w.r.t. b to get the desired result.

The above result tells us that $\Phi(az + b)\phi(z)/\Phi(b/\sqrt{1 + a^2})$ is a valid PDF. Alternately, split the range of integration from $-\infty$ to 0, and from 0 to ∞ to get

$$2 \int_{-\infty}^{\infty} \phi(x)\Phi(ax)dx = 2\left[\int_{-\infty}^{0} \phi(x)\Phi(ax)dx + \int_{0}^{\infty} \phi(x)\Phi(ax)dx\right]. \qquad (8.23)$$

Due to symmetry of the normal CDF we get $F(x) = 1 - F(-x)$, so that the second integral can be written as $\int_{-\infty}^{0} \phi(x)(1 - \Phi(ax))dx$. Adding them gives the RHS as $2 \int_{-\infty}^{0} \phi(x)dx = 1$, proving that it is indeed a PDF. This reduces to the SND defined above when $b = 0$ because $\Phi(0) = 1/2$. The skewness and kurtosis of this distribution are, respectively, $\gamma_1 = ((4 - \pi)/2)(q/p)^3$ and $\gamma_2 = 3 + 2(\pi - 3)(q/p)^4$ where $p = a/\sqrt{1 - 2a^2/(\pi(1 + a^2))}$ and $q = a\sqrt{2/(\pi(1 + a^2))}$. Extensions of this result can be found in Azzalini and Capitanio (1999) [16], Arellano-Valle, Gomez, and Quintana (2004) [10], Genton (2005) [63], etc.

Half-skew-normal distributions could also be developed on similar lines using the identities $\int_{-\infty}^{0} \phi(ax)\Phi(bx)dx = (1/(2\pi|a|)) \arctan(a/b)$, and $\int_{0}^{\infty} \phi(ax)\Phi(bx)dx = (1/(2\pi a))[\pi/2 - \arctan(a/b)]$. A representation of this in terms of the CDF of bivariate normal can be found as [106] $\int_{-\infty}^{y} \phi(ax)\Phi(a + bx)dx = \text{BivN}(a/\sqrt{a^2 + b^2}, y; \rho = -b/\sqrt{1 + b^2})$.

Problem 8.15   If $X \sim N(\mu, \sigma^2)$ then prove that $E[\Phi(x)] = \Phi(\mu/\sqrt{1 + \sigma^2})$.

The extended SN distribution has PDF

$$f(x; a, b) = C\phi(x)\Phi(a|x|)\Phi(bx), \quad \text{for } a, b, |x| < \infty, \qquad (8.24)$$

where $C = 2\pi/(\pi + 2\arctan(a))$. This reduces to SN for $a = 0$, and two-piece SN for $b = 0$. Another two-parameter skew-normal distribution reported by Bahrami, Rangin, and Rangin (2009) [21] has PDF

$$f(x; a, b) = C\phi(x)\Phi(ax)\Phi(bx), \quad \text{for } a, b, |x| < \infty, \qquad (8.25)$$

where $C = 2\pi/\cos^{-1}(-ab/[\sqrt{1 + a^2}\sqrt{1 + b^2}])$. See Adcock and Azzalini (2020) [1] for a review of skew-elliptical and related distributions.

8.5.1   PROPERTIES OF SND

Due to the symmetry of the normal CDF, the following properties of SND are immediate.

1. If $X \sim SN(a)$ then $-X \sim SN(-a)$.

2. $G(x;a) = 1-G(-x;-a)$ or equivalently $G(-x;a) = 1-G(x;-a)$ where $G()$ is the CDF of SND.

3. If $X \sim SN(a)$ then $X^2 \sim \chi_1^2$, is a central χ^2 with 1 DoF.

4. If $X \sim SN(a)$ and $U \sim N(0,1)$ is independent, then $(X + U)/\sqrt{2} \sim SN(a/\sqrt{2 + a^2})$.

5. If X_1 and X_2 are IID $N(0,1)$ random variables, then $Z = (a/\sqrt{1 - a^2})|X_1| + (1/\sqrt{1 - a^2})X_2 \sim SN(a)$.

If $G(y)$ denotes the CDF of $Y = X^2$, it follows that $G(y) = \Pr[Y \leq y] = \Pr[X^2 \leq y] = \Pr[-\sqrt{y} \leq X \leq \sqrt{y}] = F(\sqrt{y}) - F(-\sqrt{y})$ where $F()$ is the CDF of SND. Differentiate w.r.t. y to get $g(y) = \frac{1}{2\sqrt{y}}[f(\sqrt{y}) + f(-\sqrt{y})]$. Now substitute the value of the PDF to get $g(y) = \frac{2}{2\sqrt{y}}(\phi(\sqrt{y})\Phi(a\sqrt{y}) + \phi(-\sqrt{y})\Phi(-a\sqrt{y}))$. Substitute the value of $\phi(x)$ to get $g(y) = \frac{1}{\sqrt{2\pi}\sqrt{y}}\exp(-y/2)(\Phi(a\sqrt{y}) + \Phi(-a\sqrt{y}))$. Now use $\Phi(x) + \Phi(-x)=1$ to get $g(y) = \frac{1}{\sqrt{2\pi}}\exp(-y/2) = \frac{1}{2^{1/2}\Gamma(1/2)}\exp(-y/2)y^{1/2-1}$, which is the PDF of central χ^2 with 1 DoF. This shows that SN(a) is a generalized chi-distribution. More specifically, if the square of a random variable X has a χ_1^2 distribution, then X is absolutely continuous, and has a SN(a) distribution for some real a (Arnold and Lin (2004) [12]).

This result can be extended to the noncentral case when the mean of SND is nonzero as $G(y) = F(\lambda + \sqrt{y}) - F(\lambda - \sqrt{y})$ (Chattamvelli (1995b) [32]). Differentiate w.r.t. y to get the PDF of noncentral χ^2 distribution with 1 DoF. An immediate conclusion of property 3 is that the distribution of the square of the SND is independent of the skew parameter a. Similarly, the distributions of all even functions of X in general, and $|X|$ in particular are also independent of a. Thus, we can conclude that even functions of X cannot be used in goodness-of-fit tests to differentiate between normal and skewed normal distributions. An interpretation of the last item is that the realization of a process Z (like returns on investment in stocks, gains accrued in actuary, profit in finance, etc) is driven by a half-Gaussian component and an independent full-Gaussian component. Consider a pair of random variables (X, Y) (like high school GPA and SAT score) which has a bivariate normal distribution. If students admitted to one institution have "above average" scores in one of the variables (say SAT score), then the distribution of the other variable for that particular subset of students is SND.

Note that $\delta = a/\sqrt{1 - a^2}$ is an increasing function of a with $\delta = 1$ as asymptote (as a increases, the maximum value of δ is one).

Problem 8.16   If X and Y are IID $N(0,1)$ variates, and $U = (aX - Y)/\sqrt{a + a^2}$, find the conditional distribution of $X|U$.

The SN family is uni-modal. Closed form expressions for mode are complicated, but an approximate mode is $M = K - \frac{1}{2}\left(C * D + \text{sign}(a)\exp(-2\pi/|a|)\right)$ where $K = \sqrt{2/\pi}a/\sqrt{1+a^2}$, $C = \sqrt{1-K^2}$, and D is the skewness coefficient.

Problem 8.17 If X and Y are IID $Z(0,1)$ random variables. Let $U = X$ and $V = Y - \lambda X$. Prove that the conditional density of U, given $V < 0$ is $SN(\lambda)$.

Problem 8.18 If $Z \sim SND(0,1,a)$ show that $\Pr[Z < 0] = \frac{1}{2} + \frac{1}{\pi}\arctan(a)$.

Problem 8.19 If $X \sim SND(0,1,a)$ and $Z(0,1)$ is independent of X, find the distribution of $(X + Z)/\sqrt{2}$.

Skew normal distributions with location and scale parameters can be obtained from the above by a linear transformation $Y = \mu + \sigma X$ with PDF

$$f(x; \mu, \sigma, a) = (2/\sigma)\phi((x-\mu)/\sigma)\Phi(a(x-\mu)/\sigma). \tag{8.26}$$

This has mean $\mu = \sigma\delta\sqrt{2/\pi}$ where $\delta = a/\sqrt{1+a^2}$, and variance $\sigma^2(1 - 2\delta^2/\pi)$ because the second moment is $E[X^2] = 2\Phi(0) = 1$. The coefficient of variation is $\sigma/\mu = \sqrt{\pi}\sqrt{(1-2\delta^2/\pi)}/(\delta\sqrt{2})$. The "standard normalization" technique discussed in Section 8.1 in page 93 can be used to approximate the SND by N(0,1) tables by subtracting the sample mean and dividing by the sample standard deviation. This is fairly good when the skew parameter is in $(-1, +1)$. The CDF is $\Phi((x-\mu)/\sigma) - 2T((x-\mu)/\sigma, a)$. Truncated version of SND can be obtained using this expression (see Chapter 1). It has a representation as a linear combination of half-normal variable and an independent $Z(0,1)$ as follows (Henze (1986) [72]). Let U be a half-normal variate and V be an independent $Z(0,1)$ variate. Then $W = a/\sqrt{1+a^2}U + (1/\sqrt{1+a^2})V \sim SND(a)$. If $X \sim SND(a)$, the distribution of $|X|$ is independent of a and is the same as the distribution of $|Z|$ where Z is a standard normal variate.

8.5.2 THE MGF OF SND

The MGF is

$$M_x(t; a) = 2\exp(t\mu + t^2\sigma^2/2)\Phi(\sigma at/(\sqrt{1+a^2})). \tag{8.27}$$

We first find the MGF of SN(0,1). By definition $M_z(t; a) = E(\exp(tz)) = 2\int_{-\infty}^{\infty}\exp(tz)\phi(z)\Phi(az)dz$. Expand $\phi(z)$, combine with $\exp(tz)$, then complete the square to get $Mz(t; a) = 2\exp(t^2/2)\int_{-\infty}^{\infty}\frac{1}{\sqrt{2\pi}}\exp(-\frac{1}{2}(z-t)^2)\Phi(az)dz$. Put $z - t = u$ to get the RHS as $2\exp(t^2/2)\int_{-\infty}^{\infty}\frac{1}{\sqrt{2\pi}}\exp(-u^2/2)\Phi(a(u+t))du$. Write $\frac{1}{\sqrt{2\pi}}\exp(-u^2/2)$ as $\phi(u)$. Using Theorem 8.14, this reduces to $M_z(t; a) = 2\exp(t^2/2)\Phi(at/\sqrt{1+a^2})$. Expand the $\Phi()$ term to get another representation as $M_z(t; a) = 2\exp(t^2/2)\left(\frac{1}{2} + \int_0^{bt}\phi(z)dz\right)$ where $b = a/\sqrt{1+a^2}$. Differentiate w.r.t. t and put $t = 0$ to get the mean as $\mu = (2a/\sqrt{1+a^2})\phi(0) = \sqrt{2/\pi}(a/\sqrt{1+a^2})$. Differentiate a second time and put $t = 0$

to get the second moment $E(Z^2) = 2\Phi(0) = 2 \times \frac{1}{2} = 1$. From this the variance follows as $\sigma^2 = 1 - \frac{2a^2}{\pi(1+a^2)}$.

Now consider $Z = (X - \mu)/\sigma$ so that $X = \mu + \sigma Z$. Use $M_{\mu+\sigma Z}(t) = \exp(\mu t)M_z(t\sigma)$. From this we get the MGF of the general distribution $SND(\mu, \sigma, a)$ as $Mz(t; \mu, \sigma, a) = 2\exp(t\mu + t^2\sigma^2/2)\Phi(\sigma at/(\sqrt{1 + a^2}))$.

If X_1 and X_2 are IID $SND(\mu_1, \sigma, a)$ and $SND(\mu_2, \sigma, a)$ (with the same skew parameter a), then $X_1 + X_2 \sim SND(\mu_1 + \mu_2, \sigma, a)$. This follows easily from the MGF. If X_1 and X_2 are IID $SND(\mu, \sigma, a_1)$ and $SND(\mu, \sigma, a_2)$, the sum $X_1 + X_2$ is not in general SN distributed. This can be proved from the MGF because $\Phi(a_1t/(\sigma\sqrt{1 + a_1^2})\Phi(a_2t/(\sigma\sqrt{1 + a_2^2})$ cannot be expressed as a single $\Phi()$ term for arbitrary a_1 and a_2. But, if they are dependent and shares a common half-normal part, the sum is SND. More precisely, if $X_1 = b_1|Z| + \sqrt{1 - b_1^2}Z_1$ and $X_2 = b_2|Z| + \sqrt{1 - b_2^2}Z_2$ where Z_1 and Z_2 are IID $SN(0, 1, a_i)$, then $X_1 + X_2 \sim \sqrt{1 + 2b_1b_2}$ $SN((b_1 + b_2)/\sqrt{1 + 2b_1b_2})$.

The CDF has a representation in terms of Owen's T function $T(h, a) = \frac{1}{2\pi}$ $\int_0^\infty \exp(-h^2(1 - x^2)/2)/(1 + x^2)dx$ as $F(x; \mu, \sigma) = \Phi((x - \mu)/\sigma) - 2T(((x - \mu)/\sigma, a)$. As $T(h, a) = T(h, -a) = T(-h, a) = T(-h, -a)$, it follows that $G(x, -a) = \Phi(x) - 2T(x, -a)$ where $G(x, a)$ is the CDF of SND.

8.5.3 RANDOM SAMPLES

Random samples from SND are most easily generated using the decomposition of it as $X = b|Z_1| + \sqrt{1 - b^2}Z_2$ where Z_1 and Z_2 are IID standard normal variates, and $b = a/\sqrt{1 - a^2}$. A random number generator of N(0,1) can now be used to generate random numbers from SND.

8.5.4 FITTING SND

The location-and-scale SND has three parameters, namely, μ, σ, and a, which are, respectively, the mean, standard deviation, and skewness parameters. The easiest way to get estimates of the unknowns is using the method of moments. As there are three unknowns, we equate the sample mean (m), sample standard deviation (s) and sample third moment (t) to corresponding expressions for the population to get three equations in three unknowns. As the skewness $\gamma = \frac{4-\pi}{2}(D\sqrt{2/\pi})^3/(1 - 2D^2/\pi)^{3/2}$, solving for D gives $D^2 = \frac{\pi}{2}|\gamma|^{2/3}/(|\gamma|^{2/3} + (2 - \pi/2)^{2/3})$. Take the positive square root to get D, and assign the same sign as that of γ (if γ is negative, then D is also negative). Solving for the unknowns give

$$\hat{\mu} = m - \sqrt{2/\pi}\sigma D\hat{\sigma} = s/\sqrt{1 - 2D^2/\pi}\hat{a} = D/\sqrt{1 - D^2}.$$

These estimates may not be accurate for small sample sizes. Nevertheless, they can be used as starting point for MLE estimation. If sample skewness is larger than 0.99527, we cannot get a reasonable estimate of the skewness parameter a. We could skip the condition $D > 1$ in such situations and estimate a. To obtain the MLE estimates, we need to solve a set of nonlinear equations

$$\sigma^2 = \frac{1}{n} \sum_{k=1}^{n} (x_k - \mu)^2, \qquad (8.28)$$

$$a \sum_{k=1}^{n} \phi(a(x_k - \mu)/\sigma)/\Phi(a(x_k - \mu)/\sigma) = \sum_{k=1}^{n} (x_k - \mu)/\sigma$$

$$\sum_{k=1}^{n} (x_k - \mu)/\sigma \phi(a(x_k - \mu)/\sigma)/\Phi(a(x_k - \mu)/\sigma) = 0. \qquad (8.29)$$

Method of moment estimates may be used as starting point to solve this set of equations numerically. Simplified expressions ensue when the SND is reparametrized in terms of the mean μ, standard deviation σ, and asymmetry coefficient called "centered parametrization" (Azzalini and Capitano (1999) [16], Azzalini and Regoli (2012) [19]).

8.5.5   EXTENSIONS OF SND

Several extensions of the SN family exist. Examples are two-parameter SN (TSN), generalized SN (GSN), Balakrishnan SN (BSN), skew-curved SN (CSN), etc. The BSN(a) has PDF $f(x; a, n) = C\ \phi(x)[\Phi(ax)]^n$, where n is a positive integer, $C = (1/c_n(a))$ is the normalizing constant. Particular values of $c_n(a)$ are $c_1(a)=1/2$, $c_2(a) = (1/\pi)\arctan(\sqrt{1 + 2a^2})$. Several generalizations of BSN(a) also exist [10, 20, 69, 72, 133, 146]. The TSN(a, b) is the expected value of $\Phi(ax + b)$, and has PDF (Azzalini and Capitanio (2014) [17])

$$f(x; a, b) = \phi(x)\Phi(ax + b)/\Phi(b/\sqrt{1 + a^2}) \qquad (8.30)$$

discussed in page 103. This reduces to SN(a) when $b = 0$. A change-of-origin-and-scale transformation can be applied to get the 4-parameter version. If X and Y are IID Z(0,1) random variables, the distribution of $X|a + bX > Y \sim TSN(a, b)$. The GSN introduced by Arellano-Valle, Gomez, and Quintana (2004) [10] has PDF

$$f(x; a, b) = \phi(x)\Phi(ax/\sqrt{1 + bx^2}), \qquad (8.31)$$

and by Dalle-Valle has PDF

$$f(x; a, b) = \phi(x)\Phi(x)/\Phi(ax/\sqrt{1 + bx^2}). \qquad (8.32)$$

It is called skewed-curved normal (SCN) when $b = a^2$. The SCN distributions with location and scale parameters can be obtained from the above by a linear transformation $Y = \mu + \sigma x$ with PDF

$$f(x; \mu, \sigma, a, b) = 2\phi((x - \mu)/\sigma)\Phi(a(x - \mu)/\sigma)/\Phi(a(x - \mu)/\sqrt{\sigma^2 + b(x - \mu)^2}). \qquad (8.33)$$

Analogous to the relation between normal and lognormal distributions, we could also extend SND to log-SND as follows. If $\log(X)$ is distributed as $SN(a)$, then the distribution of X is called log-SND, which finds applications in data breach modeling [56]. The SN distribution has been extended to multivariate case by Azzalini and Dalla Valle (1996) [18], the PDF of which is

$$f(x; k, \Sigma, a) = 2\phi_k(x; \Sigma)\Phi(a'x), \tag{8.34}$$

where $\phi_k(x; \Sigma)$ is the multivariate normal density, a is a k-vector of constants, and $\Phi(a'x)$ is the CDF of univariate normal. This has mean vector $\mu = \sqrt{2/\pi}(1 + a'\Sigma a)^{-1/2}\Sigma a$ and variance-covariance matrix $\Sigma - \mu\mu'$. A two-piece SND introduced by Kim (2005) [84] has PDF

$$f(x; a) = 2\pi\phi(x)\Phi(a|x|)/[\pi + 2\tan^{-1}(a)]. \tag{8.35}$$

8.6   LOGNORMAL DISTRIBUTION

Log-normal distribution (LND) is a continuous distribution with range 0 to ∞. It arises in a variety of applications, especially in the size distribution of atmospheric aerosol and marine particles, size-division applications like clay, mineral and metal crunching, etc. Stock markets use it to model returns on investment in short time horizons. It is used in metallurgy and mining engineering because the grade distribution of mineral deposits is often skewed to the right. It is obtained from the UND $f(z) = \frac{1}{\sqrt{2\pi}}e^{-z^2/2}$ using the transformation $y = e^z$ or equivalently $z = \log(y)$ (the natural logarithm). This means that the transformed variate is lognormally distributed. It is important to remember that if X is normally distributed, $\log(X)$ is not lognormal (a normal variate extends from $-\infty$ to ∞, but logarithm is undefined for negative argument). This gives $\partial x/\partial y = 1/y$, so that the PDF of standard lognormal distribution becomes

$$f(y; 0, 1) = \frac{1}{\sqrt{2\pi}\,y}e^{-(\ln y)^2/2}, \; 0 \le y < \infty. \tag{8.36}$$

The two-parameter LND is easily obtained as

$$f(y; \mu, \sigma^2) = \frac{1}{\sqrt{2\pi}\,\sigma y}e^{-(\ln y - \mu)^2/(2\sigma^2)}. \tag{8.37}$$

Here, μ and σ^2 are not the mean and variance of lognormal distribution, but that of the underlying normal law (from which LND is obtained by the transformation $y = e^x$). It is generated from a sequence of multiplicative random effects. It is also called "the law of proportionate effect," "antilog normal" and "logarithmic normal" distribution.

8.6.1   ALTERNATE REPRESENTATIONS

As $1/\sqrt{2\pi} = 0.39894$, it can also be written as

$$f(y; \mu, \sigma^2) = 0.39894/(\sigma y) \quad \exp(-(\ln y - \mu)^2/(2\sigma^2)). \tag{8.38}$$

If $Y = a + b \log(X - \mu)$ has a unit normal distribution, then X is said to have a three-parameter lognormal distribution. This has PDF

$$f(x; \mu, a, b) = \frac{b}{(x-\mu)\sqrt{2\pi}\ \sigma x} \exp(-\frac{1}{2}(a + b \ln (x - \mu)^2), \quad x > \mu. \tag{8.39}$$

This provides a better fit for size distributions of atmospheric aerosol and marine particles than other distributions like exponential, inverse Gaussian, Weibull, etc.

8.6.2 PROPERTIES OF LOGNORMAL DISTRIBUTION

Gibrat's distribution is a special case of lognormal distribution for $\mu = 0$ and $\sigma = 1$. This distribution and inverse Gaussian distribution are somewhat similar-shaped for small parameter values. Tail areas can be easily evaluated using the CDF of a normal distribution. For instance, if $Y \sim \text{LND}(0,1)$ then $\Pr[Y > y_0] = \Pr[Z > \ln(y_0)] = 1 - \Phi(\ln(y_0))$. The CDF can be expressed in terms of erf() function as

$$F(x) = \frac{1}{2} \text{erfc}((\mu - \ln(x))/\sigma\sqrt{2}) = \Phi((\ln(x) - \mu)/\sigma). \tag{8.40}$$

From this, it is easy to show that the area from the mode to the mean of an LND is $\Phi(\sigma/2) - \Phi(-\sigma)$ where $\Phi()$ is the CDF of UND.

The quantiles of standard normal and lognormal are related as $Q_p(x) = \exp(\mu + \sigma Z_p(z))$ where Z_p denotes the corresponding quantile of standard normal variate (Table 8.3). Replace p by $p + 1$ and divide by the above expression to get

$$Q_{p+1}(x) = Q_p(x) \exp(\sigma (Z_{p+1}(z) - Z_p(z))). \tag{8.41}$$

The sum of several independent LNDs can be *approximated* by a scaled LND. A first-order approximation can be obtained by equating the moments of linear combination with target lognormal distribution as done by Patnaik (1949) [107] for noncentral χ^2 distribution. This is also called "moment matching method." As the cumulants of LND are more tractable, we could equate the cumulants and obtain a reasonable approximation.

Example 8.20 Mode of Lognormal Distribution Prove that LND is unimodal with the mode at $\exp(\mu - \sigma^2)$. What is the modal value?

Solution 8.21 Consider the PDF (8.39). To find the maximum we take log first, as the maximum of $f(x)$ and $\log(f(x))$ are the same. This gives $\log(f_y(\mu, \sigma^2)) = K - \log(y) - (\log y - \mu)^2/(2\sigma^2)$ where K is a constant. Differentiate w.r.t. y and equate to zero to get $-1/y - (\log y - \mu)/(y\sigma^2) = 0$. Cross-multiply and solve for y to get $(\log y - \mu) = -\sigma^2$ or equivalently $\log(y) = \mu - \sigma^2$. Exponentiate both sides to get the result $y = \exp(\mu - \sigma^2)$. Put

Table 8.3: Properties of lognormal distribution

Property	Expression	Comments
Range of X	$0 \le x < \infty$	Continuous
Mean	$\mu = \exp(\mu + \frac{1}{2}\sigma^2)$	Median $= e^\mu$
Mode	$\exp(\mu - \sigma^2)$	Mode < median < mean
Inverse moment	$E(X/\mu)^{-r} = E(X/\mu)^{r+1}$	
Variance	$e^{2\mu+\sigma^2}(e^{\sigma^2} - 1)$	$e^{2\mu}w(w-1)$ where $w = e^{\sigma^2}$
Skewness	$\gamma_1 = \sqrt{w-1}(w+2)$	Approx. symmetry as $\sigma \to 0$
Kurtosis (β_2)	$(w-1)[w^2(w+3) + 6(w+1)]$	
Mean deviation	$2e^{\mu+\sigma^2/2}[2\Phi(\sigma/2) - 1]$	$2e^{\mu+\sigma^2/2}\mathrm{erf}(\sigma/(2\sqrt{2}))$
CV	$\sqrt{(w-1)}$	
CDF	$\int_0^{\log(x)} \frac{1}{\sigma\sqrt{2\pi}} e^{-1/2((y-\mu)/\sigma)^2} dy$	$\frac{1}{2}(1 \pm P(1/2, z^2/2))$, $z = \frac{\ln x - \mu}{\sigma}$
Moments	$\mu'_k = \exp(k\mu + k^2\sigma^2/2)$	
*Log-normal (Π)	$(\mu_1, \sigma_1^2),(\mu_2, \sigma_2^2)$	$\mu_1 + \mu_2, \sigma_1^2 + \sigma_2^2$ product XY
*Log-normal (ratio)	$(\mu_1, \sigma_1^2),(\mu_2, \sigma_2^2)$	$\mu_1 + \mu_2, \sigma_1^2 + \sigma_2^2$ ratio X/Y

The product and ratio of two independent lognormal variates are log-normal with the parameters as shown. Similarly, the geometric mean of n IID lognormal variates is lognormally distributed.

the value in (8.39) to get the modal value $\frac{1}{\sqrt{2\pi}\sigma \exp(\mu-\sigma^2)} e^{-(\mu-\sigma^2-\mu)^2/(2\sigma^2)}$. This simplifies to $\frac{1}{\sqrt{2\pi}\sigma \exp(\mu-\sigma^2)} e^{-\sigma^2/2}$.

Problem 8.22   Show that the constant C in $f(x; a, b) = (C/x^2) \exp(-(a - b/x)^2/2)$ is $b/(\Phi(a)\sqrt{2\pi})$. Prove further that the mode is $b(\sqrt{a^2 + 8} - a)/4$.

Problem 8.23   Obtain the unknown C in $f(x; a, b, c, \mu) \propto C/[(x - a)(b - x)] \exp((\ln((x - a)/(b - x) - \mu)^2)/(2c))$, $a < x < b$.

8.6.3   MOMENTS

The mean is $\mu' = e^{\mu+\frac{1}{2}\sigma^2}$, and variance $\sigma'^2 = e^{2\mu+\sigma^2}(e^{\sigma^2} - 1) = e^{2\mu}\omega(\omega - 1)$ where $\omega = e^{\sigma^2}$. Whereas the variance of the general normal distribution is given by a single-scale parameter σ^2, the variance of lognormal distribution depends on both the location and scale parameters μ and σ^2. As this distribution in the "logarithmic scale" reduces to the normal law, many of the additive properties of the normal distribution have multiplicative analogues for LND. For example, the additive form of the central limit theorem asserts that the mean of a random sample tends to normality for increasing values of n can be stated for LND as follows: If $X_1, X_2, \ldots, X_n$ are independent lognormal random variables with finite $E(\log(X_i))$, then $Z = (\log(S_n) - n *$ $E[\log(X_i)])/(n * Var(\log(X_i)))^{1/2}$ asymptotically approaches normality where S_n is the product of the $X_i's$. This is the reason why it is used to model sequences of multiplicative random effects. An implication of this result is that the LND could arise either from a nonlinear transformation $(y = \exp(x))$ of normally distributed inputs or as the product of independent LND inputs.

Its third ordinary moment $\mu'_3 = \exp(3(\mu + 3\sigma^2/2))$. The skewness coefficient $\gamma_1 = (e^{\sigma^2} - 1)(2 + e^{\sigma^2})$. See Table 8.1 for further properties.

8.6.4   GENERATING FUNCTIONS

The ChF is given by

$$\phi_X(t) = \int_0^\infty \exp(ity)\frac{1}{\sqrt{2\pi}\,\sigma y}\exp(-(\ln y - \mu)^2/(2\sigma^2))dy. \tag{8.42}$$

Using a simple change-of-contour argument to convert the integral into one in which the oscillatory nature of the new integrand is devoid of the argument of the ChF, Gubner (2006) [67] obtained an efficient numerical procedure to compute it for an extremely large range of its argument (including computer programs for it).

Problem 8.24   If the lower quartile $Q_1 = 100$ and upper quartile $Q_3 = 400$ for a LND, find the mean and variance.

Problem 8.25   The check-in time for long flights at a busy airport approximately follows a lognormal distribution with parameters $\mu = 90$ min and $\sigma = 30$ min. An airline opens one more check-in counter that cuts the average wait time by half. What is the probability that a passenger has to wait more than 20 min for check-in?

Example 8.26   Mean deviation of Lognormal distribution   Find the mean deviation of LND distribution.

Solution 8.27   Let $X \sim LND(\mu, \sigma^2)$. The MD is given by

$$MD = 2\int_{ll}^{\mu'} F(x)dx = 2\int_0^c \Phi((\ln(x) - \mu)/\sigma)dx, \text{ where } c = e^{\mu + \frac{1}{2}\sigma^2}. \tag{8.43}$$

Put $z = ((\ln(x) - \mu - \sigma^2/2)$, so that $dx = e^{z+\mu+\sigma^2/2}dz$. The lower limit in (8.43) becomes $-\infty$, and the upper limit is 0 because $\ln(c) = \mu + \frac{1}{2}\sigma^2$. Thus, we get

$$MD = 2\int_{-\infty}^0 \Phi(z/\sigma + \sigma/2)e^{z+\mu+\sigma^2/2}dz = 2e^{\mu+\sigma^2/2}\int_{-\infty}^0 e^z\Phi(z/\sigma + \sigma/2)dz. \tag{8.44}$$

Take $u = \Phi(z/\sigma + \sigma/2)$, and $dv = e^z dz$ so that $v = e^z$ and $du = (1/\sigma)\phi(z/\sigma + \sigma/2)$. Apply integration by parts to (8.43) to get

$$MD = 2e^{\mu+\sigma^2/2}\left[\Phi(z/\sigma + \sigma/2)e^z \mid_{-\infty}^0 - \int_{-\infty}^0 e^z(1/\sigma)\phi(z/\sigma + \sigma/2)dz\right]. \qquad (8.45)$$

The first expression in (8.45) is $\Phi(\sigma/2)$ as $\Phi(-\infty) = e^{-\infty} = 0$. To evaluate the second expression, we use $-\int_b^a f()dx = \int_a^b f()dx$, expand $\phi()$ and write it as

$$\int_0^\infty e^z(1/\sigma)\phi(z/\sigma + \sigma/2)dz = (1/\sigma\sqrt{2\pi})\int_0^\infty \exp(z - \frac{1}{2}(z/\sigma + \sigma/2)^2)dz. \qquad (8.46)$$

Expand the quadratic, and combine the exponent as $z - \frac{1}{2}(z/\sigma + \sigma/2)^2 = -\frac{1}{2}(z^2/\sigma^2 - z + \sigma^2/4) = -\frac{1}{2}(z/\sigma - \sigma/2)^2$. This gives the above integral as

$$(1/\sigma\sqrt{2\pi})\int_0^\infty \exp(-\frac{1}{2}(z/\sigma - \sigma/2)^2)dz. \qquad (8.47)$$

Now put $(z/\sigma - \sigma/2) = v$ so that $dz = \sigma dv$. The upper limit remains the same, but the lower limit becomes $-\sigma/2$. Upon substitution, the σ cancels out from the numerator and denominator and (8.47) becomes

$$(1/\sqrt{2\pi})\int_{-\sigma/2}^\infty \exp(-v^2/2)dv = 1 - \Phi(-\sigma/2). \qquad (8.48)$$

Substitute in (8.44) to get

$$MD = 2 e^{\mu+\sigma^2/2}[\Phi(\sigma/2) + 1 - \Phi(-\sigma/2)]. \qquad (8.49)$$

Divide the area under the normal curve from $-\infty$ to $-\sigma/2$, $-\sigma/2$ to $+\sigma/2$, and from $+\sigma/2$ to $+\infty$. As the total area is unity, the expression (8.49) is simply the middle area from $-\sigma/2$ to $+\sigma/2$. Hence it becomes $2\Phi(\sigma/2) - 1$. Substitute for the bracketed expression to get

$$MD = 2 e^{\mu+\sigma^2/2}[2\Phi(\sigma/2) - 1]. \qquad (8.50)$$

Problem 8.28   The claim size for an auto-insurance is lognormal distributed with mean 285 and variance 225. Find the probability that a randomly selected claim will be greater than 200.

Example 8.29   Geometric mean of IID lognormal variates   If $X_1, X_2, \ldots, X_n$ are independent lognormal random variables $LND(\mu, \sigma^2)$, find the distribution of the $GM = (X_1 * \cdots * X_n)^{1/n}$.

Solution 8.30   As X_i is LND, $\log(X_i)$ are normally distributed. Taking log gives $Y = \log(GM) = (\log(X_1) + \cdots + \log(X_n))/n$. Each component in this expression is normal $N(\mu, \sigma^2)$ so that Y is $N(\mu, \sigma^2/n)$. Taking the inverse transformation $X = e^y$ shows that GM is lognormal, $LND(\mu, \sigma^2/n)$.

8.6.5   PARTIAL EXPECTATION OF LOGNORMAL DISTRIBUTION

The partial expectation of LND has applications in economics, finance, and insurance. It is defined as

$$g(k) = \int_{k}^{\infty} x f(x; \mu, \sigma^2) dx = e^{\mu + \sigma^2/2}[\Phi([\mu + \sigma^2 - \ln(k)]/\sigma)]. \tag{8.51}$$

Consider the survival function form of MD as

$$E|x - \mu| = 2 \int_{\mu}^{ul} S(x) dx. \tag{8.52}$$

Take $u = S(x)$, and $dv = dx$ so that $du = -f(x)$, and we get

$$E|x - \mu| = 2 \left([xS(x) \, |_{\mu}^{\infty}] + \int_{\mu}^{\infty} x \, f(x) dx \right). \tag{8.53}$$

Divide throughout by 2 and rearrange (8.53) to get

$$\int_{\mu}^{\infty} x f(x) dx = E|x - \mu|/2 + \mu S(\mu). \tag{8.54}$$

Depending on whether $k < \mu$ or $k > \mu$, the integral between them can be expressed in terms of $\Phi()$. This shows that the partial expectation of lognormal distribution is related to the MD through the SF value at μ.

8.6.6   FITTING LOGNORMAL DISTRIBUTION

We have seen above that the mean $E(X) = e^{\mu + \frac{1}{2}\sigma^2}$ and Variance $V(X) = e^{2\mu + \sigma^2}(e^{\sigma^2} - 1)$. Take log and solve for μ and σ^2 to get $\mu = \ln(E(X)) - 0.5 * \ln(1 + (Var(X)/E(X)^2))$, and $\sigma^2 = \ln(1 + (Var(X)/E(X)^2))$. If the sample size is sufficiently large, we could replace $E(X)$ by the sample mean $\overline{x}_n$, and $Var(X)$ by s_n^2, and obtain estimates of the unknown parameters.

8.7   APPLICATIONS

The normal distribution has been used for decades as the standard for several experimental procedures and analyses. The primary reasons for its popularity are the simplicity, malleability, and availability of extensive tables. It is extensively used for model building in physical, engineering and natural sciences. The normal distribution is also used to model situations that give rise to sums of independent random variables. The central limit theorem states that "a scaled sum of n independent processes (such as measurement errors) approach the normal law as n becomes large." This is why physical quantities that are expected to be the sum of many independent measurements are assumed to be nearly normal. It is written in opto-electronics in the form

$$p(N, \Delta t) = (1/(\sqrt{2\pi \overline{N}})) \exp(-\Delta N^2/(2\overline{N})), \tag{8.55}$$

where N is the number of particles (electrons in light-detecting diodes or photons in photonics), $\overline{N}$ the average in a unit time interval Δt, so that $\overline{N} = a \Delta t$.

Skewed data occurs in many practical experiments. The SND is a prime candidate to be considered when empirical distribution of a variable resembles a normal law with slight asymmetry. One example is the distribution of BMI among college students, which is almost always positively skewed. Reaction times in experimental psychology, marks of students in difficult exams, file sizes on hard disks, and settlement amounts in insurance claims are often positively skewed.

González-Val (2019) [66] found that the size distribution of nearby U.S. cities is lognormally distributed when distance separation is less than 100 miles. Rare-earth elements and radioactivity, micro-organisms in closed boundary regions, solute mobility in plant cuticles, pesticide distribution in farm lands, time between infection, and appearance of symptoms in certain diseases like COVID-19 follow approximately the LND. Singer (2013) [135] observes that metal contents of mineral deposits are not lognormally distributed, but when grouped into 28 deposit types, over 90% are lognormal. It also has applications in insurance and economics (Crow and Shimuzu (2018), [50, Chapter 9]) and ecological economics (Shanmugam (2017) [127, 128]). It is the widely used parametric model in mining engineering for low-concentration mineral deposits, reliability engineering of the distributions for the coefficients of friction and wear-and-tear (Steele (2008) [138]), digital wireless applications for outages (Beaulieu, et al. (1994) [25]), and wireless communications (Cardieri and Rappaport (2000) [29]). Heat conduction through semi-infinite solids in which no heat is self-generated (within it) can be modeled using *error function* as $[(T_e - T)/(T_e - T_0)]\mathrm{erf}(x/(2\sqrt{ct}))$ where T_0 is the boundary condition at time $t = 0$ (initial temperature), T_e is the boundary condition at time $x = 0$ (maximum attainable temperature on the boundary at some saturation interval $t > 0$), and c is a solid-specific constant. It is assumed that a constant temperature is maintained at boundary layers. Doppler shape in photonics and ultrasonics is assumed to be Gaussian distributed as

$$f(x; a, b, C) = (C/(b\sqrt{\pi})) \quad \exp(-[(x - a)/b]^2), \tag{8.56}$$

where a denotes the line center, $b = (a/c)\sqrt{2KT/m}$, m is the molecular mass, K is Boltzman constant, T is the absolute temperature in $K°$, and c is the speed of light. This is called Lorentz distribution in optics.

Several absolute heights (called asperity-height) in engineering fields are assumed to be Gaussian distributed. One-dimensional Brownian motion of a molecule at infinite dilution in a uniform solvent can be approximated by the Gaussian law as

$$f(z, t) = (K\pi)^{-1/2} \exp(-z^2/K), \quad K = 2v\lambda^2 t, \tag{8.57}$$

where v is the number of steps of size λ in one time unit, and $f(z, t)$ is the probability of finding the molecule between z and $z + dz$ at time t if it starts at $t = 0$. The probability of finding electrons with velocity v in electrical conductors, and the distribution of heat-current when a uniform charge is applied at one end is approximately Gaussian distributed, where the scale parameter (spread σ) depends on the presence of heat in the medium. Similarly, the diffusion profile through thin-film coatings is similar in form to a scaled Gaussian law

$$f(x,t) = (M/\sqrt{2})(K\pi)^{-1/2}\exp(-x^2/K), \quad K = 4Dt, \tag{8.58}$$

where M is the total amount of diffusant per unit area, D is the diffusion coefficient, t is the time of diffusion, and $f(x,t)$ is the concentration of the diffusant at distance x from the boundary (Note that this can be converted into a probability distribution by dividing the LHS by $M/\sqrt{2}$).

The production of oil-and-gas (OG) projects starts out at small volume and increases steadily until it reaches the production plateau, for which the production curve can be approximated by LND (Cheng, et al. (2019) [43]). The lognormal distribution is also used to model stock prices when continuously compounded returns of a stock follow normal distribution. In other words, it arises in multiplicative growth processes. Most of the financial asset return distributions are non-normal, for which suitable alternatives are LND and SND. The Black–Scholes model used in price options and price movements in finance uses the LND as well.

8.7.1 PROCESS CONTROL

Data are assumed to be normally distributed in process control applications. One typical example is in SQC for which control limits are set using normal quantiles. In practice, these data deviate slightly from a Gaussian law. In some manufacturing environments, it is also noticed that moderate to strong asymmetry occurs due to wear and tear of machines and parts thereof. Setting control limits based on the UND may not be the best option as it could result in wrong decisions and loss of revenues. If the exact nature of skewness is known, we could build process control charts using quantiles of SND instead that performs far better than UND based control charts. See Shanmugam (1998) for an index method of determining the sample size for normal populations.

8.7.2 CLASSIFICATION

The classical linear discriminant analysis (LDA), introduced by Fisher, assumes that data are approximately normally distributed. Similarly, the scaled Mahalanobis distance used in multi-class classification has a noncentral F distribution when the population is normal. This assumption does not hold in thermal engineering, data communications, satellite imagery, and several fields in actuarial sciences like property liability insurance, and in management like finance, sales and stock markets. Simple data transformation techniques or half-tail modeling techniques can be used when departures from normality are minimal. A highly successful and useful technique uses the skew-normal model or its extensions in such situations. For example, images obtained by remote sensing satellites are divided into pixels, and the intensity distributions are assumed to be Gaussian. This distribution is often skewed due to a multitude of reasons. Image classification techniques will suffer from misclassification errors in such situations. In other words, ignoring the skewness of data results in decreased efficiency and wrong conclusions. The skew-normal discriminant analysis (SNDA) introduced by Azzalini and Capitanio (1999) [16] is an extension of LDA when skewness and covariance matrices are equal, but means are unequal.

8.8 SUMMARY

The normal distribution and its extensions are introduced in this chapter, and their basic properties discussed. It is the most extensively studied distribution of continuous family as it finds applications in a wide variety of fields. This distribution is of special interest to theoretical statisticians due to central limit theorem. There are several statistical distributions that tends to the normal law for special parameter values. In addition, an extension of it called skew-normal family has gained keen interest among modelers during the last three decades because several practical data inevitably deviate from a perfectly bell-shaped curve.

CHAPTER 9

Cauchy Distribution

9.1   INTRODUCTION

The Cauchy distribution is named after the French mathematician Augustin L. Cauchy (1789–1857), although it was known to Pierre De Fermat (1601–1665) and Isaac Newton (1642–1727), much earlier. It is a symmetric, unimodal, and continuous distribution with PDF

$$f(x; a, b) = 1/\left[b\pi(1 + (x-a)^2/b^2)\right] \quad a, b > 0, \ -\infty < x < \infty. \tag{9.1}$$

The location parameter is "a" and scale parameter is "b". The Standard Cauchy Distribution (SCD) is obtained from the above by putting $a = 0$ and $b = 1$ as

$$f(x) = \frac{1}{\pi} \frac{1}{1 + x^2}, \quad -\infty < x < \infty. \tag{9.2}$$

It is called the Lorentz distribution, Cauchy–Lorentz distribution or Breit–Wigner distribution in particle physics, where the PDF is written as

$$f(u, \rho, u_0) \propto \frac{\rho}{\pi u_0} \frac{1}{1 + (u/u_0)^2}, \quad -\infty < u < \infty, \tag{9.3}$$

in which u is the potential (like gravitational potential) and u_0 is the initial value.[1] The parameter u_0 (b in (9.1)) is half the Full Width at Half Maximum (FWHM) in optics, laser communications and spectrometry. For example, the Lorentzian spectral line shape in optics is described by a wrapped Cauchy law (wrapped Lorentzian distribution) in which a is the resonance frequency, and b is half the FWHM line-width.

9.1.1   ALTERNATE REPRESENTATIONS

The PDF of SCD is sometimes expressed as

$$f(x) = (1/\pi)(x^2 + 1)^{-1} = \left(\pi(1 + x^2)\right)^{-1}, \ 0 < |x| < \infty. \tag{9.4}$$

As $1/\pi = 0.31831$, (9.4) can also be written as

$$f(x) = 0.31831(1 + x^2)^{-1} = 0.31831/(x^2 + 1), \ 0 < |x| < \infty. \tag{9.5}$$

---

[1] ρ is a scale-factor of the density, and is not a part of the PDF itself.

The corresponding representation of general case is

$$f(x; a, b) = (b/\pi) \left([b^2 + (x - a)^2]\right)^{-1} = (b\pi)^{-1} \left([1 + (x - a)^2/b^2]\right)^{-1} \quad a, b > 0, \quad I_{-\infty,\infty}, \tag{9.6}$$

where $I()$ denotes the indicator function. As shown below, a is the median and b is the quartile deviation. Write $b^2 + (x - a)^2 = (b - i(x - a))(b + i(x - a))$ (where i is the imaginary constant) and split using partial fractions to get another representation.

McCullagh's parametrization, introduced by Peter McCullagh (1992) [100], uses a single complex-valued parameter $\theta = a + ib$ where $i = \sqrt{-1}$ is the imaginary complex number and extends the range of b to include negative values. This, denoted by MCauchy(a, b), has PDF

$$f(x; a, b) = (|b|\pi)^{-1} \left(1 + (x - a)^2/b^2\right)^{-1} \quad a > 0, \quad -\infty < x < \infty. \tag{9.7}$$

This is considered to be degenerate for $b = 0$. This can be written in terms of θ as

$$f(x; \theta) = (I(|\theta|)\pi)^{-1}|x - \theta|^{-2} \quad a > 0, \quad -\infty < x < \infty, \tag{9.8}$$

where $I()$ denotes the imaginary part. If $X \sim$ MCauchy(a, b), then $Y = (c_1 X + d_1)/(c_2 X + d_2)$ is distributed as Cauchy($c_1\theta + d_1, c_2\theta + d_2$) where c_1, c_2, d_1, d_2 are real numbers.

A transformation to polar coordinates using $a = \rho \sin(\theta)$ and $b = \rho \cos(\theta)$ results in the PDF

$$f(x; \rho, \theta) = \rho \cos(\theta)/[\pi(x^2 - 2\rho \sin(\theta)x + \rho^2)], \quad 0 < \rho < \infty, |\theta| < \pi/2. \tag{9.9}$$

If $(R(t), \theta(t))$ denotes the coordinates at time t of a random point in standard Brownian motion, then $(\theta(t) - \theta(0)\mathrm{mod}(2\pi))/\log(t)$ approaches a Cauchy distribution as $t \to \infty$ if $R(0) > 0$ (Spitzer (1958) [136]).

It is easy to prove that X/ρ and ρ/X are both Cauchy distributed, which follows from the fact that if U is CUNI(0,1) distributed, then so is $1/U$ (Williams (1969) [142]). A circular Cauchy distribution (also called wrapped Cauchy distribution) is defined on a unit circle, and has PDF

$$f(x; w) = \frac{1}{2\pi} \frac{1 - |w|^2}{|\exp(ix) - w|^2}, \quad x \in [0, 2\pi). \tag{9.10}$$

Problem 9.1 If $X \sim$ Cauchy($-1, 1$), find the distribution of $Y = (X - (1/X))/2$.

9.2 PROPERTIES OF CAUCHY DISTRIBUTION

This distribution is symmetric around the location parameter with more probability farther out in the tails than the normal distribution (Figure 9.1). That is why it is sometimes used to model data containing outliers. The Cauchy distribution is closed under the formation of sums of independent random variables, but does not obey the central limit theorem which

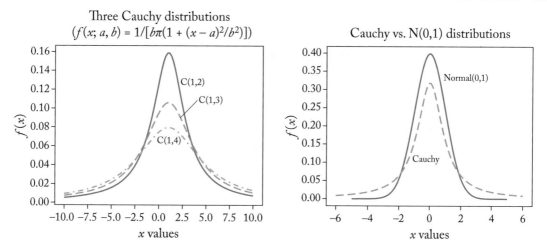

Figure 9.1: Cauchy distributions.

states that the arithmetic mean of n IID random variables approaches the normal law in the limit (but the harmonic mean of n IID continuous random variables with $p(x) > 0$ approaches the Cauchy law in the limit if X possesses the first derivative at $x = 0$). Since the integral $\int_{-\infty}^{\infty} x\,dx/(1 + x^2)$ does not exist, the mean is undefined, but fractional and absolute moments exist. Thus, $E(|x|^k) = \Gamma((1 + k)/2)\Gamma((1 - k)/2)/\pi = 1/\cos(\pi k/2)$ for k real or complex such that $|Re(k)| < 1$. This has the indirect implication that the mean of a random sample will keep on increasing as the sample size becomes larger and larger (due to the heavier tails of this distribution). Obviously, the probability of event occurrences in the tails are greater than corresponding event occurrences of normal law. The same is the case with variance. The limiting value $\frac{1}{\pi} \underset{R \to \infty}{Lt} \int_{-R}^{R} \frac{x}{1+x^2} dx$ is zero.

Problem 9.2 Prove that the first and third quartiles of general Cauchy distribution are $a \mp b$. What are the quartiles of MCauchy(a, b)?

The characteristic function is

$$\phi(t; a, b) = \frac{1}{b\pi} \int_{-\infty}^{\infty} \frac{e^{itx}}{1 + ((x - a)/b)^2} dx = \frac{2}{b\pi} \int_{0}^{\infty} \frac{\cos(tx)}{1 + ((x - a)/b)^2} dx = \exp(iat - b|t|). \tag{9.11}$$

Although higher-order moments do not exist (as per the definition of moments as integrals), one could argue that the odd moments are all zeros because the PDF is symmetric around the location parameter. Median and mode of the general Cauchy distribution coincide at $x = a$, with modal value $1/(b\pi)$ (Table 9.1). The median remains the same for symmetrically truncated Cauchy distribution. The points of inflection of the general PDF are at $a \mp b/\sqrt{3}$. Whereas the modal value of SCD is 0.318309886, that of N(0,1) is 0.39894228 (the difference being 0.0806).

Table 9.1: Properties of Cauchy distribution $(1/\left[b\pi[1+(x-a)^2/b^2]\right])$

Property	Expression	Comments
Range of X	$-\infty < x < \infty$	Continuous; infinite
Mean	μ does not exist	
Median	a	Mode = a
Variance	σ^2 = does not exist	
Skewness	$\gamma_1 = 0$	Symmetric
Kurtosis	$\beta_2 = 9$	Leptokurtic
Mean deviation	Does not exist	
CDF	$\frac{1}{2} + \frac{1}{\pi}\tan^{-1}((x-a)/b)$	
Moments	Does not exist	
$Q_1 = a - b$	$Q_3 = a + b$	
ChF	$\exp(ita - |t|b)$	

The modal values are approximately equal for $N(0, \sigma^2 = \pi/2 = 1.570796)$ (for $\sigma \approx 1.2533$). As the quartiles are $Q_1 = a - b$ and $Q_3 = a + b$ we get $\Pr[X < a] = 0.50$, $\Pr[a - b < X < a] = \Pr[a < X < a + b] = 0.25$. The CDF of SCD is

$$F(x; a, b) = 1/2 + (1/\pi) \tan^{-1} x, \qquad (9.12)$$

and that of general Cauchy distribution is given by $F_x(a, b) =$

$$\frac{1}{\pi} \int_{-\infty}^{x} \frac{dy}{[b\pi[1 + (y-a)^2/b^2]]} = \frac{1}{2} + \text{sign(x-a)}\ (1/\pi)\ \tan^{-1}\left(\frac{x-a}{b}\right). \qquad (9.13)$$

From this the inverse CDF follows as $F^{-1}(u) = a + b[\tan(\pi(u - 0.5))]$. As $1/\sqrt{(3)} = 0.57735$, the points of inflection are well within the quartiles. The hazard function of SCD is $h(x) = 1/[(1 + x^2)(\pi/2 - \arctan(x))]$, and the cumulative hazard function is $H(x) = -\ln(\pi/2 - \arctan(x)/\pi)$. This distribution is infinitely divisible. Thus, the PDF can be expressed as an n-fold convolution of itself.

9.3   RELATION WITH OTHER DISTRIBUTIONS

It is a special case of Student's t distribution when the degree of freedom is 1 (this fact can be used for random sample generation from Cauchy distribution). The ratio of IID random variables occurs in ranking and selection problems as well as in physical sciences. It is well known that the ratio X/Y of two IID $N(0,1)$ variates is standard Cauchy distributed, but the converse need not be true (Laha (1959) [92]). This result holds even when the denominator (Y) is distributed as half-normal. The ratio of the absolute value of two IID $N(0,1)$ variates (or equivalently the ratio of two IID folded $N(0,1)$ variates) has the standard half-Cauchy distribution (HCD) with PDF $f(x) = \frac{2}{\pi}\frac{1}{1+x^2}$ and SF $S(x) = 1 - \frac{2}{\pi}\tan^{-1}(x)$. The PDF and CDF of the scaled version is $f(x;a) = \frac{2}{\pi}\frac{a}{a^2+x^2}$ and $F(x;a) = \frac{2}{\pi}\tan^{-1}(x/a)$. Several researchers have also extended it to the ratio of IID skew-normal (Azzalini type) random variables (Behboodian (2006) [26], Huang and Chen (2007) [73]). The HCD is a special case of folded Student's t distribution with one degree of freedom. Let X_1 and X_2 be IID $N(0,1)$ random variables, and k be a constant. Define $Y = |X_1|^k$, and $Z = |X_2|^k$. Then the ratio $W = Y/Z = |X_1/X_2|^k$ has a generalized Cauchy distribution with PDF

$$f(w,k) = 2/[|k|\, B(.5,.5)]w^{1/k-1}/(1 + w^{2/k}), \quad w > 0, \tag{9.14}$$

where $B(.5,.5) = \Gamma(.5)^2/\Gamma(1) = \pi$ (Rao and Garg (1969) [112]). This reduces to the HCD for $k = \pm 1$. Many other generalizations of Cauchy distributions are available in the literature. For example,

$$f(x;a,b,\sigma,p) = b\sigma(\sigma^p + |x-a|^p)^{-2/p}, \quad w > 0, \tag{9.15}$$

where $b = p\Gamma(2/p)/[2\Gamma(1/p))^2$ (Rider (1957) [117], Carrillo, Aysal, and Barner (2010) [30]). Here, a is the location parameter, b is the scale-parameter, and p is called the tail-parameter (which is any real number). This reduces to the Cauchy distribution for $p = 2$. The tails decay slower than that of Cauchy distribution for $p < 2$, resulting in a much heavier-tailed distribution.

If Z is Cauchy distributed, then $(c + dZ)$ is Cauchy distributed where c, d are real constants. Many other similar relationships can be found in Arnold (1979) [11] and Norton (1983) [104]. As the t-distribution with 1 DoF is the standard Cauchy distribution, we can define a skew Cauchy distribution as a special case of skew-t distribution (for $n = 1$) as $T = SN(a)/Y$ where $Y \sim \chi_n^2/n$ where $SN(a)$ is the skew normal distribution and Y is an independent scaled χ^2 distribution. This is used in distorted risk modeling in finance, differential pricing models in stock markets, etc.

The inverse-Cauchy distribution is the distribution of $Y = 1/X$. More precisely, if $X \sim$ Cauchy$(a.b)$, then $Y = 1/X$ is distributed as Cauchy(a',b') where $a' = a/D$, $b' = b/D$ and $D = a^2 + b^2$. It is also related to the standard bivariate normal distribution. If $X \sim U(0,1)$, then $\tan(\pi(x - 0.5))$ is standard Cauchy distributed. Alternately, if $X \sim U(-\pi/2, \pi/2)$ then $Y = \tan(X)$ is Cauchy distributed. This fact is used to generate random numbers from a Cauchy distribution.

Although the double exponential (Laplace) distribution is more peaked than Cauchy distribution, the latter has heavier tails than Cauchy distribution.

Example 9.3 Distribution of integer part   If X has Cauchy distribution, find the distribution of the integer part $Y = \lfloor X \rfloor$.

Solution 9.4   We have $f(x) = 1/[\pi(1 + x^2)]$ for $-\infty < x < \infty$. The random variable Y takes integer values on the entire real line including $y = \mp\infty$. Specifically,

$$\Pr[Y = y] = P[y \le X < y + 1] = \int_y^{y+1} 1/[\pi(1 + x^2)]dx = (1/\pi)\tan^{-1}(x)|_y^{y+1}$$

$$= (1/\pi)[\tan^{-1}(y + 1) - \tan^{-1}(y)].$$

A similar expression could be obtained for $x < 0$ as $\Pr[Y = y] = \Pr[y - 1 \le X < y]$. As alternate terms in above equation cancel out, this form is useful to compute the CDF, and to prove that the probabilities add up to 1. For example, sum from $-\infty$ to ∞ to get $\tan(\infty) - \tan(-\infty) = \pi/2 - (-\pi/2) = \pi$, which cancels with π in the denominator. Now use the identity $\tan^{-1}(x) - \tan^{-1}(y) = \tan^{-1}((x - y)/(1 + xy))$ to get

$$\Pr[Y = y] = (1/\pi)\tan^{-1}(y + 1 - y)/[1 + y(y + 1)] = (1/\pi)\tan^{-1}(1/[1 + y(y + 1)]), \tag{9.16}$$

which is the desired form.

9.3.1   FUNCTIONS OF CAUCHY VARIATES

If X is Cauchy distributed, then so is its reciprocal $Y = 1/X$. If $X \sim \text{Cauchy}(a,b)$ then X/b and b/X are both Cauchy distributed. If X_1, X_2 are IID Cauchy distributed random variables with parameters a_1, b_1 and a_2, b_2, then $X \pm Y \sim \text{Cauchy}(a_1 \pm a_2, b_1 \pm b_2)$. In particular, if $X_1 \sim \text{Cauchy}(0, b_1)$ and $X_2 \sim \text{Cauchy}(0, b_2)$ are independent, then $X_1 + X_2 \sim \text{Cauchy}(0, b_1 + b_2)$. If $X_1, X_2, \ldots X_n$ are independent Cauchy distributed random variables, then $\overline{X} = (X_1 + X_2 + \cdots + X_n)/n$ is also Cauchy distributed. This is most easily proved using the MGF or ChF. An implication of this result is that the Cauchy mean does not obey the central limit theorem (CLT) or the law of large numbers (the law of large numbers states that S_n/n for SCD converges to μ in probability, and the CLT states that the distribution of S_n/n tends to $N(0,1)$ as $n \to \infty$). This is primarily due to the fact that more of the Cauchy distribution tails are sampled as sample size n becomes large, so that the diminution in the variance of $\overline{x}$ given by σ^2/n is overtaken by the outliers like sample values in the extreme tails. Any convex combination of IID standard Cauchy random variables also has a Cauchy distribution. Special truncated distributions can be obtained either using symmetric truncations at $-c$ and $+c$ or asymmetric truncations in both tails, as shown below. A special case is the Cauchy distribution with range $(-1, +1)$ with PDF

$$f(x; a, b) = 1/[\pi[\tan^{-1}((1 - a)/b) + \tan^{-1}((1 + a)/b)](b^2 + (x - a)^2)]. \tag{9.17}$$

Problem 9.5   If X is standard Cauchy distributed, then prove that $Y = X/\sqrt{1 + X^2}$ has PDF

$$g(y) = (1/\pi)(1 - y^2)/(1 - y^2)^{3/2} = 1/[\pi (1 - y^2)^{1/2}], \quad -1 \le y \le +1. \tag{9.18}$$

As noted above, the Cauchy distribution is related to the uniform distribution $U(0,1)$ through the tangent function. If $X \sim U(-\pi/2, \pi/2)$, then $Y = a + b\tan(X)$ has a

Cauchy(a, b) distribution. In addition, $Z = \tan(nX)$ where n is a positive integer is also Cauchy distributed. This allows us to find tangent functions of Cauchy distributed variates like $2X/(1 - X^2)$, $(3X - X^3)/(1 - 3X^2)$, $4(X - X^3)/(1 - 6X^2 + X^4)$, etc., which are, respectively, $\tan(2X)$, $\tan(3X)$, $\tan(4X)$, etc. (Pittman and Williams (1967) [110], Kotlarski (1977) [88]). These are all Cauchy distributed. If X and Y are IID Cauchy variates, $(X - Y)/(1 - XY)$ and $(X - 1)/(X + 1)$ are identically distributed. In addition, Y and $(X + Y)/(1 - XY)$ are independent (Arnold (1979) [11]).

Problem 9.6   If $X \sim$ Cauchy$(0,1)$, find the distribution of $Y = (1/X - X)/2$. Hint: Use $2\cot(2u) = \cot(u) - \tan(u)$.

Problem 9.7   If U_1 and U_2 are IID CUNI$(-\pi/2, \pi/2)$ random variables, and $Y_1 = \tan(U_1)$, $Y_2 = \tan(U_2)$, $Y_3 = -\tan(U_1 + U_2)$, then prove that Y_1, Y_2, and Y_3 are pairwise independent SCD variates, but are jointly dependent.

9.3.2   PARAMETER ESTIMATION

The mean is no better than a single observation in estimating the unknown population parameters. A practical technique is to take the median of a large sample as the estimator of the location parameter (a) and half the sample inter-quartile range as an estimator of scale parameter (b). As the quartiles are $Q_1 = a - b$ and $Q_3 = a + b$ with sum $Q_1 + Q_3 = 2a$, we could estimate the location parameter (a) as half the sum of the estimates of the quartiles. As $(Q_3 - Q_1)/2 = b$, the estimate of scale parameter of general Cauchy distribution is half the inter-quartile range (namely, the quartile deviation). In general, if x_p and x_{1-p} are any two *quantiles*, an unbiased estimator of the location parameter is $\hat{a} = (x_p + x_{1-p})/2$, and the estimator of the scale parameter is $\hat{b} = (x_p - x_{1-p})/2 \;\; \tan(\pi q)$ where $q = 1 - p$. Other possibilities are to use an average of a few symmetrically weighted quantiles or a linear combination of order statistics. The *method-of-moments* estimators of complete (un-truncated) Cauchy distribution are meaningless as the moments do not exist (although it could be used for double-truncated Cauchy distribution).

The likelihood functions for MLE are given by

$$\partial \log L/\partial a = \sum_{k=1}^{n} 2(x_k - a)/[b^2 + (x_k - a)^2] = 0,$$

$$\partial \log L/\partial b = n/b - \sum_{k=1}^{n} 2b/[b^2 + (x_k - a)^2] = 0. \tag{9.19}$$

Rearrange the second equation and divide by n to get

$$\frac{2}{n} \sum_{k=1}^{n} 1/\left(1 + [(x_k - a)/b]^2\right) = 1. \tag{9.20}$$

Similarly, the first equation gives

$$\sum_{k=1}^{n}[(x_k - a)/b]/[1 + [(x_k - a)/b]^2] = 0. \tag{9.21}$$

Split the numerator into two terms and use the first equation to simplify this to the form

$$\frac{2}{n}\sum_{k=1}^{n}x_k/[1 + [(x_k - a)/b]^2] = a. \tag{9.22}$$

Since the distribution tails off slowly, data values from extreme tails are a possibility. This means that even a single term can dominate the likelihood function. As a is continuous and unknown, the likelihood could change sign as we maximize w.r.t. a in the neighborhood of the largest sample value. The MLE estimator of scale parameter b is asymptotically optimal, and can be used to narrow down the search for the global optimum of the estimator of location parameter a (because there could be multiple optima for the first equation).

9.3.3 TAIL AREAS

Left-tail area of a Cauchy distribution up to c is given by

$$F(x;a,b,c) = \frac{1}{\pi}\int_{-\infty}^{c}\frac{b}{b^2 + (x-a)^2}dx, \tag{9.23}$$

which upon the substitution $y = (x - a)/b$ becomes $F(x;a,b,c) = \frac{1}{\pi}$ $(\tan^{-1}((c - a)/b) + \pi/2)$. When $c = -d$ is negative, this becomes $F(x;a,b,d) = \frac{1}{\pi}$ $(\pi/2 - \tan^{-1}((d + a)/b))$. A similar expression for right-tail area from $c > 0$ to ∞ follows as $S(x;a,b,c) = \frac{1}{\pi}(\pi/2 - \tan^{-1}((c - a)/b))$. Put $a = 0$, $b = 1$ to get the corresponding expressions for SCD.

Problem 9.8 Prove that the average of n IID Cauchy random variables has PDF $f(x) = n/[\pi(n^2 + x^2)]$.

Problem 9.9 If $c > 1$ is a real constant, prove that $\int_{-t}^{ct} xdx/[\pi(1 + x^2)] = \frac{1}{2\pi}\ln(\frac{1+c^2t^2}{1+t^2})$, which approaches $\ln(c)/\pi$ as $t \to \infty$. (Hint: split the integral from $-t$ to t and t to ct, and use property of odd integral in the first expression.)

9.4 SPECIAL CAUCHY DISTRIBUTIONS

Several variants of the Cauchy distributions exist. Examples are truncated, skewed, and transmuted Cauchy distributions. As the moments of this distribution does not exist, size-biased Cauchy distribution is meaningless, although we could form size-biased truncated Cauchy distributions, and absolute-moment based size-biased Cauchy distribution. As $E(|x|^k) = 1/\cos(\pi k/2)$ for k such that $|Re(k)| < 1$, a size-biased Cauchy distribution can be defined as

$$f(x;k) = \frac{\cos(\pi k/2)}{\pi} \frac{|x|^k}{1 + x^2}, \quad -\infty < x < \infty. \tag{9.24}$$

9.4.1 TRUNCATED CAUCHY DISTRIBUTION

As the Cauchy distribution tails off slowly compared to the normal law, it has been a suitable choice for modeling in electrical and mechanical engineering, calibration systems, marine biology, mineral geology, etc. Due to the limited range of respective variable in these fields, the truncated Cauchy distribution is used for modeling. One example is to model the spread of diseases in plants in water and species on the move (except humans who travel a lot) (see page 131). The distribution of the arithmetic mean of n IID Cauchy variates is independent of n. This has the implication that the sample mean is not a good estimator of the location parameter. Nevertheless, this does not apply to the mean of IID truncated Cauchy variates. The truncation can occur either in one of the tails or in both tails (called double truncation). A question that naturally arises is how much truncation should be present for the mean to be used as an estimator of the location parameter. As it is the tail of this distribution that is problematic, truncation in both tails (say beyond 10th and 90th percentiles) will result in a distribution with moments having finite values. In the presence of a sample of size n, the left truncation point must satisfy $c \leq x_{(1)}$ and the right truncation point must satisfy $d \geq x_{(n)}$. Left-truncated Cauchy distribution is obtained by truncating at an integer c. The resulting PDF is

$$f(x;a,b,c) = \frac{1}{K\pi} \frac{b}{b^2 + (x-a)^2}, \quad c < x < \infty, \tag{9.25}$$

where $K = \int_c^\infty \frac{1}{\pi} \frac{b}{b^2+(x-a)^2} dx$. Put $y = (x-a)/b$ and $dx = bdy$ to obtain $K = \frac{1}{\pi}(\pi/2 - \tan^{-1}((c-a)/b))$. The π cancels out from numerator and denominator to get

$$f(x;c) = \frac{1}{\pi/2 - \tan^{-1}((c-a)/b)} \frac{b}{b^2 + (x-a)^2}, \quad c < x < \infty. \tag{9.26}$$

Put $a = 0, b = 1$ to get the left-truncated SCD as

$$f(x;c) = \frac{1}{\pi/2 - \tan^{-1}(c)} \frac{1}{1 + x^2}, \quad c < x < \infty. \tag{9.27}$$

The zero-truncated Cauchy(a,b) (ZTC) distribution is a special case. As the CDF of Cauchy(a,b) is $\frac{1}{\pi}\left(\tan^{-1}((x-a)/b) + \pi/2\right)$, the PDF of ZTC is $f(x;a,b)/[1 - F(0;a,b)]$. Put the values to get the PDF of ZTC as

$$f(x;a,b) = \frac{b}{b^2 + (x-a)^2} \frac{1}{\pi/2 + \tan^{-1}(a/b)}, \quad 0 < x < \infty. \tag{9.28}$$

The half-Cauchy distribution results when truncation is up to the median (a). If the distribution is truncated in both tails at $x = \mp c$, the resulting PDF is $\frac{1}{2\tan^{-1}(c)} \frac{1}{1+x^2}$ for $-c \le x \le +c$. This has mean 0 and variance $c/\tan^{-1}(c) - 1$.

Now consider a general Cauchy distribution with location parameter a and scale parameter b with PDF in (9.5). An asymmetrically truncated Cauchy distribution is obtained by truncating at c and $d > c$. This has PDF

$$f(x; a, b, c, d) = \frac{1}{K\pi} \frac{b}{b^2 + (x-a)^2}, \quad c < x < d, \tag{9.29}$$

where the normalizing constant K is given by $K = \int_c^d \frac{1}{\pi} \frac{b}{b^2+(x-a)^2} dx$. Put $u = (x-a)/b$ so that $du = dx/b$. The limits of u are $(c-a)/b$ and $(d-a)/b$. Thus, $K = \int_{(c-a)/b}^{(d-a)/b} \frac{1}{\pi} \frac{1}{1+u^2} b \, du$ $= \frac{1}{\pi} \tan^{-1}(u)|_{(c-a)/b}^{(d-a)/b} = \frac{1}{\pi}[\tan^{-1}((d-a)/b) - \tan^{-1}((c-a)/b)]$. The $1/\pi$ cancels out from numerator and denominator resulting in the PDF

$$f(x; a, b, c, d) = b[\tan^{-1}((d-a)/b) - \tan^{-1}((c-a)/b)]^{-1} / (b^2 + (x-a)^2), \quad c < x < d. \tag{9.30}$$

This has CDF

$$(\tan^{-1}((x-a)/b) - \tan^{-1}((c-a)/b))$$
$$/(\tan^{-1}((d-a)/b) - \tan^{-1}((c-a)/b)), \quad c < x < d. \tag{9.31}$$

The mean is easily found as $\mu = a + \frac{b}{2} \frac{\log((b^2+(d-a)^2)/(b^2+(c-a)^2))}{\tan^{-1}((d-a)/b) - \tan^{-1}((c-a)/b)}$. Using $\log(u/v) = \log(u) - \log(v)$, this can also be expressed as $\mu = a + \frac{b}{2} \frac{\log(1+((d-a)/b)^2) - \log((1+((c-a)/b)^2))}{\tan^{-1}((d-a)/b) - \tan^{-1}((c-a)/b)}$. In the case of symmetric truncation, c is in the left-tail and d is in the right-tail such that $c = -d$. As $\tan^{-1}((-d-a)/b) = -\tan^{-1}((d+a)/b)$, the PDF becomes

$$f(x; a, b, c) = [\tan^{-1}((c-a)/b) + \tan^{-1}((c+a)/b)]^{-1} \frac{b}{b^2 + (x-a)^2} \quad -c < x < c. \tag{9.32}$$

This has mean a. A recurrence relation for ordinary moments can easily be obtained as

$$\mu'_n = \frac{b}{(n-1)K}(d^{n-1} - c^{n-1}) + 2a\mu'_{n-1} - (a^2 + b^2)\mu'_{n-2}, \tag{9.33}$$

where $K = \tan^{-1}((d-a)/b) - \tan^{-1}((c-a)/b)$. From this the variance can be obtained as $V(X) = E(X^2) - [E(x)]^2 = b((d-c)/K - b) - (a - \mu)^2$, which shows that the variance is maximum in the case of symmetric truncation. If the truncation points are expressed as an offset from the location parameter a, this becomes

$$f(x; a, b, c) = [2\tan^{-1}(c/b)]^{-1} \frac{b}{b^2 + (x-a)^2} \quad a - c < x < a + c. \tag{9.34}$$

This distribution has moments of all orders. In particular, the mean is a and the variance is $bc/\tan^{-1}(c/b) - b^2$.

Problem 9.10 Obtain the PDF for a symmetrically both-side truncated Cauchy distribution with the truncation point θ. Does the variance exist?

Problem 9.11 If X is distributed as double truncated Cauchy(a, b, c, d) where c and d denote the lower and upper truncation points, find the distribution of $Y = \tan^{-1}(X - a)/b$.

Problem 9.12 If $U \sim \text{CUNI}(a, b)$, find the distribution of $Y = a + b\tan(U)$.

9.4.2 LOG-CAUCHY DISTRIBUTION

A random variable has log-Cauchy distribution if $Y = \log(X)$ has Cauchy distribution. In other words, if Y has a Cauchy distribution, then $X = \exp(Y)$ has a log-Cauchy distribution. It belongs to the log-location-and-scale family of distributions [81]. It is called a "super heavy-tailed" distribution because the tailing off to the right is slower than that of Pareto distribution. The PDF is given by

$$f(x; a, b) = b(x\pi)^{-1}\left([b^2 + (\log(x) - a)^2]\right)^{-1}$$
$$= 1/[b\pi x \ (1 + ((\log(x) - a)/b)^2)] \quad a, b > 0, \ x > 0. \tag{9.35}$$

This has median $\exp(a)$. The standard log-Cauchy distribution is obtained when $a = 0$ and $b = 1$. This has PDF

$$f(x; 0, 1) = (x\pi)^{-1}\left([1 + (\log(x))^2]\right)^{-1}; \quad x > 0 \tag{9.36}$$

and CDF and SF are

$$F(x; 0, 1) = \frac{1}{2} + \frac{1}{\pi}\arctan(\ln x) \quad S(x) = \frac{1}{2} - \frac{1}{\pi}\arctan(\ln x); \quad x > 0, \tag{9.37}$$

where ln denotes logarithm to the base e. This distribution is used to model survival processes where significant extreme results could occur as in epidemiology where people become ill after infected with coronavirus, and in pharmacology to model pharmacokinetic data (Lindsey, Jones, and Jarvis (2001) [94]). Extreme events like annual maximum one-day rainfalls and river discharges can also be modeled.

Example 9.13 HM of Cauchy variates Prove that the HM of n IID Cauchy random variables has PDF f(x)=$n/[\pi(n^2 + x^2)]$.

Solution 9.14 We have seen above that if X is Cauchy distributed, then $1/X$ is identically distributed. As the HM is $n/(1/X_1 + 1/X_2 + \cdots + 1/X_n)$ the denominator expression is Cauchy distributed so that n/Cauchy is also Cauchy distributed. The result follows from the independence assumption.

Example 9.15   Functions of Cauchy variates   If X and Y are independent Cauchy distributed, prove that (i) $Z = (X + Y)/2$ and (ii) $(X - Y)/(1 + XY)$ are also Cauchy distributed.

Solution 9.16   We know that the ChF of Cauchy distribution is exp(-$|it|$). To find the ChF of $Z = (X + Y)/2$, first use $M_{cx}(t) = M_x(tc)$ where $c = 1/2$, and then use $M_{x+y}(t) = M_x(t) * M_y(t)$ to get $M_z(t) = \exp(-|it/2|) * \exp(-|it/2|) = \exp(-|it|)$, which is the ChF of Cauchy distribution. For part (ii) put $X = \tan(U)$, $Y = \tan(V)$ so that $(X - Y)/(1 + XY) = [\tan(U) - \tan(V)]/[1 + \tan(U) * \tan(V)] = \tan(U - V)$. As $X = \tan(U)$, $U = \tan^{-1}(X)$ is uniform distributed, and so is $V = \tan^{-1}(Y)$.

9.4.3   HALF-CAUCHY DISTRIBUTION

This is very similar to the half-normal distribution discussed in the last chapter where one half of the symmetric range is discarded. The half-Cauchy distribution has PDF $f(x) = 2/[\pi(1 + x^2)]$ for $x > 0$. The general form is $(1/\pi b)2/((1 + (x - a)^2/b^2))$ for $x \geq a$. This has CDF $F(y) = (2/\pi)\arctan((x - a)/b)$ for $x \geq a$. This distribution is right-skewed with median $a + b$. As the moments of right-truncated distributions exist, we could define size-biased half-Cauchy distributions, as discussed in Sections 1.3.

9.4.4   SKEW-CAUCHY DISTRIBUTION

The ratio of a standard SND over an independent $Z(0,1)$ has a Cauchy distribution. A skew Cauchy distribution (S-Cauchy) results when the denominator is taken as the absolute value of standard normal. Skewed-truncated Cauchy distribution has PDF $f(x) = 2g(x)G(ax)$, where $g()$ and $G()$ are the PDF and CDF of a truncated Cauchy distribution (Nadarajah, Kotz (2007) [103]). Behboodian, Jamalizadeh, and Balakrishnan (2006) [26] obtained the PDF of S-Cauchy distribution as

$$f(x; a) = \frac{1}{\pi(1 + x^2)}\left(1 + ax/\sqrt{1 + (1 + a^2)x^2}\right), \quad \text{for } a, |x| < \infty, \tag{9.38}$$

where $C = 2\pi/\cos^{-1}(-ab/[\sqrt{1 + a^2}\sqrt{1 + b^2}]))$.

Many other generalizations of the Cauchy distribution exist.

9.4.5   BETA-CAUCHY DISTRIBUTION

The beta-Cauchy distribution (BCD) is obtained by replacing $F(x)$ in the following expression with the CDF of one of the Cauchy distributions discussed above. It has CDF

$$G(x; p, q) = (1/B(p, q))\int_0^{F(x)} t^{p-1}(1 - t)^{q-1}dt, \text{, for } 0 < p, q, x < \infty, \tag{9.39}$$

where $B(p,q)$ is the complete beta function (Alzaatreh, Famoye & Lee(2013) [8]). The kth raw moment of this distribution exists when both p, q are greater than k.

9.5   APPLICATIONS

The size distribution of changes in the gross domestic product (GDP) is approximated by the Laplace distribution in the central part, and Cauchy distribution in the tails for almost all countries (Williams, et al. (2017) [143]). It is used to model processes resulting from spinning objects and homogeneous broadening processes as in the shape of spectral lines in spectroscopy. The limiting distribution of velocity differences in turbulent flows and separation velocity of 3D-singular vortexes in aerodynamics are approximated by a Cauchy distribution (Min, Mezic, and Leonard (1996) [101]). A Cauchy distribution describes the energy width of a particular state in quantum mechanics when the state decay exponentially over time. It is customary to express the quadratic term in the denominator of Cauchy law as a norm in these fields to get vorticity norm distribution as

$$f(x;a,b) = \frac{1}{\pi} \frac{b}{b^2 + |x - a||x - a|'}, \quad 0 < x < \infty, \tag{9.40}$$

where $|x - a|$ is the velocity difference. The infrared absorption of single spectrally-isolated molecule has an observed shape that can be approximated by the Cauchy distribution as

$$f(x;a,b,C) = (C/\pi) \quad b/[b^2 + (x - a)^2], \tag{9.41}$$

where a denotes the line center, b is the half width at half maximum, and C the line intensity in g^{-1} cm. This is called Lorentz distribution in optics. Similarly, the homogeneous broadening of radiation in photon emission occurring in the range $[v, v + dv]$ is expressed as

$$f(v; v_0, \gamma) = (\gamma/\pi)/[(v - v_0)^2 + \gamma^2], \tag{9.42}$$

where γ is the HWHM and v_0 is the central value. It is used in nuclear physics and quantum mechanics to model the energy of an unstable particle state, and in ballistic heat conduction studies. Electrical engineers use it to model the imaginary part of complex electrical permittivity. The variation in contact resistivity of resistance scatter from resistor to resistor is approximately Cauchy distributed (Winterton, Smy, and Tarr (1992) [144]).

The Voigt-profile of Fabry–Perot interferometers (also called etalon in optics, spectrometry, and crystallography[2]) used in precise length measurements (optometry) and electromagnetic wave filtering by detecting Raman-spectrum uses a convolution of Gaussian and Cauchy laws as $V(x, \sigma, b) = \int_{t=-\infty}^{\infty} g(t, \sigma) c(x - t, b) dt$ where $g(t, \sigma)$ is the PDF of N(0,σ) and $c(x - t, b)$ is

---

[2]Etalon is a French word, meaning "a measuring gauge" or "standard." Technically, a Fabry–Perot interferometer and a Fabry–Perot etalon are different because the distance between the reflecting plates are tunable in the former, so that the wavelengths at which transmission peaks occur can be changed.

the PDF of Cauchy($0,b$) distribution, which has an alternate representation in terms of complementary error function as $V(x, \sigma, b) = \mathrm{Re}(F(z))/(\sigma\sqrt{2\pi})$ where $F(z) = \exp(-z^2)\mathrm{erfc}(-iz)$, $z = (x + ib)/(\sigma\sqrt{2})$ (Golabczak and Konstantynowicz (2009) [65]). The transmission function in Fabry–Perot interferometers (used in optics, spectrometry, and astronomy (e.g., gravitational wave detection)) can be written as an infinite series of Lorentzian (Cauchy) functions [74]. Heavy-tailed noise characteristics of color image processing can be modeled using the Cauchy law as well. Altunbasak and Kamaci (2004) [5] observes that the distribution of Discrete Cosine Transform (DCT) coefficients in video coding systems like H.264 codec is closer to a Cauchy law than to a Laplacian law.

It is known as the Breit–Wigner distribution in particle physics. Physicists use it to describe the energy spectrum close to a resonance as

$$f(E, a, b) = \frac{1}{b\pi} \frac{1}{1 + ((E - a)/b)^2}, \tag{9.43}$$

where $a =\; < E >$ is the particle mass, $b = \Gamma/2$ is half the width of the resonance, and E is the energy. Using Fourier transform, it can be proved that the mean lifetime of the resonance particle is inversely proportional to the width of the resonance peak.

9.5.1 THERMODYNAMICS

Heat transfer is an important field in thermodynamics. The heat transfer coefficient (HTC) is the ratio of heat flux and temperature difference both of which could fluctuate about zero. A Cauchy distribution gives a good approximation to the heat transfer coefficient h as

$$f(x; h_0, b) = \frac{1}{\pi b}/[1 + ((x - h_0)/b)^2], \tag{9.44}$$

where h_0 is the location and b is the scale parameter.

9.5.2 CARBON NANOTUBE STRAIN SENSING

Carbon Nanotubes (CNT) are used as strain sensors due to its spectral properties, malleability, and low cost. The G'-band Raman shift (in the Raman spectrum) of CNT is sensitive to axial deformations. The spectrum curve can be modeled by a Cauchy law as

$$f(x; \omega_{(\theta)}, \lambda) = (1/\pi) \quad \lambda/[\lambda^2 + (x - \omega_{(\theta)})^2], \tag{9.45}$$

where θ is the angle of the CNT axis direction, $\omega_{(\theta)}$ is the G'-band Raman shift, and λ is the half-width at half-maximum. As the sum of IID Cauchy variates are Cauchy distributed, the spectrum information of multiple independent uniformly dispersed CNT passing around a solid body can be modeled by the Cauchy law.

9.5.3   BIOLOGY

Consider a plant that reproduces using seeds. Let X denote the distance by which the seeds are separated from the mother plant (without human intervention) by pure natural causes like winds, water currents, birds, and animals, or landscape characteristics like elevation and slope. There are a large number of seeds for which the distribution of X is heavy tailed (this depends on many factors like floating capacity of seeds in the air (as in cotton) or water (like apple and coconut), height of the plant, size, and shape of the seed and fruit, etc. as well as external factors like germination conditions, immercibility in soil, etc.). Seeds can spread in all directions in unconstrained landscape. Thus, a bivariate distribution is ideal for the modeling for land-based plants. A heavy-tailed distribution like the Cauchy law can be used to check the spread in a fixed radial direction. This technique can also be used for modeling the spread of water-borne plants and creatures. Depending on whether there are tidal movements (in both directions) or uni-directional flow, we could model it as univariate Cauchy assuming that around the point of observation, the path of the river or canal is more or less straight. A skewed Cauchy fit may be more appropriate when either the flow is in one direction (as in rivulets running down a mountain or discharge canals) or the upstream and downstream tides differ in velocity.

9.6   SUMMARY

This chapter introduced the Cauchy distributions, and its basic properties. Special Cauchy distributions like truncated, skewed, and log-Cauchy distributions are briefly introduced. The chapter ends with a list of applications of Cauchy distribution in engineering and scientific fields.

Bibliography

[1] Adcock, C. and Azzalini, A. (2020). A selective overview of skew-elliptical and related distributions and of their applications, *Symmetry*, 12:118. DOI: 10.3390/sym12010118. 103

[2] Ahsanullah, M., Shakil, M., and Golam-Kibria, B. M. (2019). On a generalized raised cosine distribution: Some properties, characterizations and applications, *Moroccan Journal of Pure and Applied Analysis*, 5(1):63–85. DOI: 10.2478/mjpaa-2019-0006. 87, 88

[3] Aigner, D. J. and Lovell, C. A. K. (1977). Formulation and estimation of stochastic frontier production function model, *Journal of Econometrics*, 12:21–37. DOI: 10.1016/0304-4076(77)90052-5. 101

[4] Aljarrah, M. A., Lee, C., and Famoye, F. (2014). On generating tx family of distributions using quantile functions, *Journal of Statistical Distributions and Applications*, 1:1–17. DOI: 10.1186/2195-5832-1-2. 9

[5] Altunbasak, Y. and Kamaci, N. (2004). An analysis of the DCT coefficient distribution with the H.264 video coder, *IEEE International Conference on Acoustics, Speech, and Signal Processing*, Montreal, 17–21. DOI: 10.1109/icassp.2004.1326510. 130

[6] Alzaatreh, A., Lee, C., and Famoye, F. (2013). A new method for generating families of continuous distributions, *Metron*, 71(1):63–79. DOI: 10.1007/s40300-013-0007-y. 9

[7] Alzaatreh, A., Lee, C., Famoye, F., and Ghosh, I. (2016). The generalized Cauchy family of distributions with applications, *Journal of Statistical Distributions and Applications*, 3(12). https://link.springer.com/article/10.1186/s40488-016-0050-3 DOI: 10.1186/s40488-016-0050-3.

[8] Alzaatreh, A., Famoye, F., and Lee, C. M. S. (2013). Beta-Cauchy distribution: Some properties and applications, *Journal of Statistical Theory and Applications*, 12(4):378–391. DOI: 10.2991/jsta.2013.12.4.5. 129

[9] Amburn, S. A., Lang, A. S., and Buonaiuto, M. A. (2015). Precipitation forecasting with gamma distribution models for gridded precipitation events in eastern Oklahoma and northwest Arkansas, *Weather Forecasting*, 30(2):349–367. https://journals.ametsoc.org/waf/article/30/2/349/39869 DOI: 10.1175/waf-d-14-00054.1. 37

[10] Arellano-Valle, R. B., Gomez, H. W., and Quintana, F. A. (2004). A new class of skew-normal distribution, *Communications in Statistics—Theory and Methods*, 33:1465–1480. DOI: 10.1081/sta-120037254. 9, 103, 107

[11] Arnold, B. C. (1979). Some characterizations of the Cauchy distribution, *Australian Journal of Statistics*, 21(2):166–169. DOI: 10.1111/j.1467-842x.1979.tb01132.x. 121, 123

[12] Arnold, B. C. and Lin, G. D. (2004). Characterizations of the skew-normal and generalized chi distributions, *Sankhya*, 66(4):593–606. 104

[13] Ashu, V., Arjun, T., and Ram, K. (2017). Optimal allocation of distributed solar photovoltaic generation in electrical distribution system under uncertainties, *Journal of Electrical Engineering and Technology*, 12(4):1386–1396. DOI: 10.5370/JEET.2017.12.4.1386. 53

[14] Azzalini, A. (1985). A class of distributions which includes the normal ones, *Scandinavian Journal of Statistics*, 12:171–178. 9, 101

[15] Azzalini, A. (1986). Further results on a class of distributions which includes the normal ones, *Statistica*, 46:199–208. 101

[16] Azzalini, A. and Capitanio, A. (1999). Statistical application of the multivariate skew-normal distribution, *Journal of Royal Statistical Society-B*, 65:367–389. DOI: 10.1111/1467-9868.00194. 103, 107, 116

[17] Azzalini, A. and Capitanio, A. (2014). *The Skew-Normal and Related Families*, Cambridge University Press. DOI: 10.1017/cbo9781139248891. 107

[18] Azzalini, A. and Dallavalle, A. (1996). The multivariate skew-normal distribution, *Biometrika*, 83(4):715–726. DOI: 10.1093/biomet/83.4.715. 108

[19] Azzalini, A. and Regoli, G. (2012). Some properties of skew-symmetric distributions, *Annals of the Institute of Statistical Mathematics*, 64:857–879. DOI: 10.1007/s10463-011-0338-5. 107

[20] Bahrami, W., Agahi, H., and Rangin, H. (2009). A two-parameter Balakrishnan skew-normal distribution, *Journal of Statistical Research*, 6:231–242. DOI: 10.18869/acad-pub.jsri.6.2.231. 107

[21] Bahrami, W., Agahi, H., and Rangin, H. (2009). A two-parameter generalized skew-Cauchy distribution, *Journal of Statistical Research*, 7:61–72. DOI: 10.18869/acad-pub.jsri.7.1.61. 103

[22] Balakrishnan, N. and Nevzorov, V. B. (2003). *A Primer on Statistical Distributions*, John Wiley, NY. DOI: 10.1002/0471722227. 43, 60

[23] Baldyga, J., Makowski, L., and Orciuch, W. (2005). Interaction between mixing, chemical reactions, and precipitation, *Industrial Engineering and Chemical Research*, 44(14):5342–5352. DOI: 10.1021/ie049165x. 53

[24] Barato, A. C., Roldan, E., Martinez, I. A., and Pigolotti, S. (2018). Arcsine laws in stochastic thermodynamics, *Physical Review Letters*, 121. https://journals.aps.org/prl/pdf/10.1103/PhysRevLett.121.090601 DOI: 10.1103/physrevlett.121.090601. 66, 67

[25] Beaulieu, N. C., Abu-Dayya, A. A., and McLane, P. J. (1994). Comparison of methods of computing lognormal sum distributions and outages for digital wireless applications, *Proc. of IEEE ICC*, pages 1270–1275. DOI: 10.1109/icc.1994.368899. 114

[26] Behboodian, J., Jamalizadeh, A., and Balakrishnan, N. (2006). A new class of skew-Cauchy distributions, *Statistics and Probability Letters*, 76:1488–1493. DOI: 10.1016/j.spl.2006.03.008. 9, 121, 128

[27] Beirlant, J., de Waal, D. J., and Teugels, J. L. (2000). A multivariate generalized Burr-Gamma distribution, *South African Statistics Journal*, 34(2):111–133. 77

[28] Bogachev, L. and Ratanov, N. (2011). Occupation time distributions for the telegraph process, *Stochastic Processes and Their Applications*, 121(8):1816–1844, DOI: 10.1016/j.spa.2011.03.016. 66

[29] Cardieri, P. and Rappaport, P. S. (2000). Statistics of the sum of lognormal variables in wireless communications, *Proc. of IEEE Vehicular Technology Conference*, 3:1823–1827. DOI: 10.1109/vetecs.2000.851587. 114

[30] Carrillo, R. E., Aysal, T. C., and Barner, K. E. (2010). A generalized Cauchy distribution framework for problems requiring robust behavior, *EURASIP Journal on Advances in Signal Processing*, 312989. DOI: 10.1155/2010/312989. 121

[31] Chattamvelli, R. (1995a). A note on the noncentral beta distribution function, *The American Statistician*, 49:231–234. DOI: 10.2307/2684647. 47, 51, 52

[32] Chattamvelli, R. (1995b). Another derivation of two algorithms for the noncentral χ^2 and F distributions, *Journal of Statistical Computation and Simulation*, 49:207–214. DOI: 10.1080/00949659408811572. 104

[33] Chattamvelli, R. (1996). On the doubly noncentral F distribution, *Computational Statistics and Data Analysis*, 20(5):481–489. DOI: 10.1016/0167-9473(94)00054-m. 51

[34] Chattamvelli, R. (2010). Power of the power-laws and an application to the PageRank metric, *PMU Journal of Computing Science and Engineering*, PMU, Thanjavur, TN 613403, India, 1(2):1–7. 28

[35] Chattamvelli, R. (2011). *Data Mining Algorithms*, Alpha Science, Oxford, UK. 15

[36] Chattamvelli, R. (2012). *Statistical Algorithms*, Alpha Science, Oxford, UK. 9, 49, 51

[37] Chattamvelli, R. (2016). *Data Mining Methods*, 2nd ed., Alpha Science, UK.

[38] Chattamvelli, R. and Jones, M. C. (1995). Recurrence relations for noncentral density, distribution functions, and inverse moments, *Journal of Statistical Computation and Simulation*, 52(3):289–299. DOI: 10.1080/00949659508811679. 46

[39] Chattamvelli, R. and Shanmugam, R. (1995). Efficient computation of the noncentral χ^2 distribution, *Communications in Statistics—Simulation and Computation*, 24(3):675–689. DOI: 10.1080/03610919508813266.

[40] Chattamvelli, R. and Shanmugam, R. (1998). Computing the noncentral beta distribution function, Algorithm AS-310, *Applied Statistics, Royal Statistical Society*, 41:146–156.

[41] Chattamvelli, R. and Shanmugam, R. (2019). *Generating Functions in Engineering and the Applied Sciences*, Morgan & Claypool. DOI: 10.2200/s00942ed1v01y201907eng037.

[42] Chattamvelli, R. and Shanmugam, R. (2020). *Discrete Distributions in Engineering and the Applied Sciences*, Morgan & Claypool. DOI: 10.2200/s01013ed1v01y202005mas034. 21, 68

[43] Cheng, C., Wang, Z., Liu, M., and Ren, X. (2019). Risk measurement of international oil and gas projects based on the value at risk method, *Petroleum Science*, 16:199–216, Springer. https://link.springer.com/article/10.1007/s12182-018-0279-1 DOI: 10.1007/s12182-018-0279-1. 115

[44] Cohen, A. C. and Whitten, B. J. (1986). Modified moment estimation for the three-parameter gamma distribution, *Journal of Quality Technology*, 17:1470–154. DOI: 10.1080/00224065.1986.11978985. 76

[45] Coifman, R. R. and Steinerberger, S. (2019). A remark on the arcsine distribution and the Hilbert transform, *Journal of Fourier Analysis and Applications*, 25:2690–2696. DOI: 10.1007/s00041-019-09678-w. 67

[46] Cordeiro, G. M. and Lemonte, A. J. (2014). The McDonald arcsine distribution: A new model to proportional data, *Statistics*, 48:182–199. DOI: 10.1080/02331888.2012.704633. 64

[47] Cordeiro, G. M., Lemonte, A. J., and Campelo, A. K. (2016). Extended arcsine distribution to proportional data: Properties and applications, *Studia Scientiarum Mathematicarum Hungarica*, 53:440–466. DOI: 10.1556/012.2016.53.4.1346. 64

[48] Craig, C. C. (1941). Note on the distribution of noncentral t with an application, *Annals of Mathematical Statistics*, 17:193–194. DOI: 10.1214/aoms/1177731752. 51

[49] Cribari-Neto, F. and Vasconcellos, K. L. P. (2002). Nearly unbiased maximum likelihood estimation for the beta distribution, *Journal of Statistical Computation and Simulation*, 72(2):107–118. DOI: 10.1080/00949650212144. 52

[50] Crow, E. L. and Shimuzu, K. (Eds.) (2018). *Lognormal Distributions—Theory and Applications*, CRC Press, FL. DOI: 10.1201/9780203748664. 114

[51] de Waal, D. J., van Gelder, P. H. A. J. M, and Beirlant, J. (2004). Joint modeling of daily maximum wind strengths through the multivariate Burr-Gamma distribution, *Journal of Wind Engineering and Industrial Aerodynamics*, 92:1025–1037. DOI: 10.1016/j.jweia.2004.06.001. 77

[52] Ding, C. G. (1996). On the computation of the distribution of the square of the sample multiple correlation coefficient, *Computational Statistics and Data Analysis*, 22:345–350. DOI: 10.1016/0167-9473(96)00002-3. 51

[53] Ding, C. G. and Bargmann, R. E. (1991). Sampling distribution of the square of the sample multiple correlation coefficient, *Applied Statistics*, 41:478–482. 51

[54] Dong, J. (2009). Mean deviation method for fuzzy multi-sensor object recognition, *IEEE International Conference*, Nanchang, pages 201–204. DOI: 10.1109/icmecg.2009.75. 9

[55] Dutka, J. (1981). The incomplete beta function—a historic profile, *Archive for History of Exact Sciences*, 24:11–29. DOI: 10.1007/bf00327713. 41

[56] Edwards, B., Hofmeyr, S., and Forrest, S. (2016). Hype and heavy tails: A closer look at data breaches, *Journal of Cybersecurity*, 2(1):3–14. DOI: 10.1093/cybsec/tyw003. 108

[57] Ellison, B. E. (1964). Two theorems for inference about the normal distribution with applications in acceptance sampling, *Journal of American Statistical Association*, 64:89–95. DOI: 10.1080/01621459.1964.10480702. 94

[58] Erel, E. and Ghosh, J. B. (2011). Minimizing weighted mean absolute deviation of job completion times from their weighted mean, *Applied Mathematics and Computation*, pages 9340–9350. DOI: 10.1016/j.amc.2011.04.020. 9

[59] Eugene, N., Lee, C., and Famoye, F. (2002). Beta-normal distribution and its applications, *Communications in Statistics—Theory and Methods*, 31(4):497–512. DOI: 10.1081/sta-120003130. 9

[60] Feller, W. (1966). *An Introduction to Probability Theory and its Applications*, John Wiley, NY. 4

[61] Fisher, R. A. (1934). The effects of methods of ascertainment upon the estimation of frequencies, *Annals of Eugenics*, 6:13–25. DOI: 10.1111/j.1469-1809.1934.tb02105.x. 6

[62] Frankl, P. and Maehara, H. (1990). Some geometric applications of the beta distribution, *Annals of the Institute of Statistical Mathematics*, Springer, 42(3):463–474. https://link.springer.com/article/10.1007/BF00049302 DOI: 10.1007/bf00049302. 53

[63] Genton, M. C. (2005). Discussion of the skew-normal, *Scandinavian Journal of Statistics*, 32(2):189–198, DOI: 10.1111/j.1467-9469.2005.00427.x. 103

[64] Glaser, R. E. (1980). A characterization of Bartlett's statistic involving incomplete beta functions, *Biometrika*, 67(1):53–58. DOI: 10.1093/biomet/67.1.53. 50

[65] Golabczak, M. and Konstantynowicz, A. (2009). Raman-spectra evaluation of the carbon layers with Voigt profile, *Journal of Achievements in Materials and Manufacturing Engineering*, 37(2):270–276. 130

[66] González-Val, R. (2019). Lognormal city size distribution and distance, *Economics Letters*, 181(8):7–10. DOI: 10.1016/j.econlet.2019.04.026. 114

[67] Gubner, J. A. (2006). A new formula for lognormal characteristic functions, *IEEE Transactions on Vehicular Technology*, 55(5):1668–1671. DOI: 10.1109/tvt.2006.878610. 111

[68] Guo, X., Jarrow, R. A., and de Larrard, A. (2014). The economic default time and the arcsine law, *Journal of Financial Engineering*, 1(3):1450025 DOI: 10.1142/s2345768614500251. 67

[69] Gupta, R. C. and Gupta, R. D. (2004). Generalized skew-normal model, *Test*, 13:501–524. DOI: 10.1007/bf02595784. 107

[70] Gupta, R. D. and Kundu, D. (2009). A new class of weighted exponential distributions, *Statistics*, 43(6):621–634. DOI: 10.1080/02331880802605346. 9

[71] Haight, F. A. (1965). On the effect of removing persons with N or more accidents from an accident prone population, *Biometrika*, 52:298–300. DOI: 10.1093/biomet/52.1-2.298. 50

[72] Henze, N. A. (1986). A probabilistic representation of the skew-normal distribution, *Scandinavian Journal of Statistics*, 13:271–275. https://www.jstor.org/stable/4616036 105, 107

[73] Huang, W.-J. and Chen, Y.-H. (2007). Generalized skew-Cauchy distribution, *Statistics and Probability Letters*, 77:1137–1147. DOI: 10.1016/j.spl.2007.02.006. 121

[74] Ismail, N., Kores, C. C., Geskus, D., and Pollnau, M. (2016). Fabry-Pérot resonator: Spectral line shapes, generic and related airy distributions, line-widths, finesses, and performance at low or frequency-dependent reflectivity, *Optics Express*, 24(15):16366–16389. DOI: 10.1364/OE.24.016366. 130

[75] Jambunathan, M. V. (1954). Some properties of beta and gamma distributions, *Annals of Mathematical Statistics*, 25:401–405. DOI: 10.1214/aoms/1177728800. 94

[76] Jiang, J. J., Ping, H., and Fang, K. T. (2015). An interesting property of the arcsine distribution and its applications, *Statistics and Probability Letters*, 105(20):88–95. DOI: 10.1016/j.spl.2015.06.002. 67

[77] Jogesh-Babu, G. and Rao, C. R. (1992). Expansions for statistics involving the mean absolute deviation, *Annals of the Institute of Statistical Mathematics*, 44:387–403. DOI: 10.1007/bf00058648. 9

[78] Johnson, D. (1997). The triangular distribution as a proxy for the beta distribution in risk analysis, *Journal of the Royal Statistical Society D, (The Statistician)*, 46(3):387–398. DOI: 10.1111/1467-9884.00091. 41

[79] Johnson, N. L. (1957). A note on the mean deviation of the binomial distributions, *Biometrika*, 44:532–533. DOI: 10.1093/biomet/44.3-4.532. 10

[80] Jones, M. C. (2015). Univariate continuous distributions: Symmetries and transformations, *Journal of Statistical Planning and Inference*, 161:119–124. DOI: 10.1016/j.jspi.2014.12.011. 7

[81] Jones, M. C. and Angela, N. (2015). Log-location-scale-log-concave distributions for survival and reliability analysis, *Journal of Statistics*, 9(2):2732–2750. DOI: 10.1214/15-ejs1089. 7, 127

[82] Jones, M. C. and Balakrishnan, N. (2003). How are moments and moments of spacings related to distribution functions?, *Journal of Statistical Planning and Inference*, 103:377–390. DOI: 10.1016/s0378-3758(01)00232-4. 14

[83] Kamat, A. R. (1965). A property of the mean deviation for a class of continuous distributions, *Biometrika*, 52:288–289. DOI: 10.1093/biomet/52.1-2.288. 10

[84] Kim, H. J. (2005). On a class of two-piece skew-normal distributions, *Statistics*, 39(6):537–553. DOI: 10.1080/02331880500366027. 108

[85] Kipping, D. M. (2013). Parametrizing the exoplanet eccentricity distribution with the Beta distribution, *Monthly Notices of the Royal Astronomical Society: Letters*, 434(1):L51–L55. DOI: 10.1093/mnrasl/slt075.

[86] Kono, H. and Koshizuka, T. (2005). Mean absolute deviation model, *IEEE Transactions*, 37:893–900. DOI: 10.1080/07408170591007786. 9

[87] Kokonendji, C. and Khoudar, M. (2004). On strict arcsine distribution, *Communications in Statistics—Theory and Methods*, 33(5):993–1006. DOI: 10.1081/sta-120029820. 67

[88] Kotlarski, I. I. (1977). An exercise involving Cauchy random variables, *The American Mathematical Monthly*, 86:229. 123

[89] Kozar, I., Malic, N. T., and Rukavina, T. (2018). Inverse model for pull-out determination of steel fibers, *Coupled Systems Mechanics*, 7(2):197–209. DOI: 10.12989/csm.2018.7.2.197. 66

[90] Krysicki, W. (1999). On some new properties of the beta distribution, *Statistics and Probability Letters*, 42(2):131–137. DOI: 10.1016/s0167-7152(98)00197-7. 43, 60

[91] Küchler, U. and Tappe, S. (2008). Bilateral gamma distributions and processes in financial mathematics, *Stochastic Processes and Their Applications*, 118(2):261–283. DOI: 10.1016/j.spa.2007.04.006. 77

[92] Laha, R. G. (1959). On a class of distribution functions where the quotient follows the Cauchy law, *Transactions of the American Mathematical Society*, 93:205–215. DOI: 10.1090/s0002-9947-1959-0117770-3. 121

[93] Lemonte, A. J., Cordeiro, G. M., and Moreno-Arenas, G. (2016). A new useful three-parameter extension of the exponential distribution, *Statistics*, 50:312–337. DOI: 10.1080/02331888.2015.1095190. 9

[94] Lindsey, J. K., Jones, B., and Jarvis, P. (2001). Some statistical issues in modelling pharmacokinetic data, *Statistics in Medicine*, 20(17–18):2775–278. DOI: 10.1002/sim.742. 127

[95] Lorek, P., Los ,G., Gotfryd, K., and Zagørski, F. (2020). On testing pseudorandom generators via statistical tests based on the arcsine law, *Journal of Computational and Applied Mathematics*, 380. DOI: 10.1016/j.cam.2020.112968. 67

[96] Mahdavi, A. and Kundu, D. (2017). A new method for generating distributions with an application to exponential distribution, *Communications in Statistics—Theory and Methods*, 46(13):6543–6557. DOI: 10.1080/03610926.2015.1130839. 9

[97] McDonald, J. B. (1995). A generalization of the beta distribution with applications, *Journal of Econometrics*, 66(1–2):133–152. DOI: 10.1016/0304-4076(94)01612-4. 9

[98] Marsaglia, M. (1961). Generating exponential random variables, *Annals of Mathematical Statistics*, 32:899–900. DOI: 10.1214/aoms/1177704984. 28, 36

[99] Marshall, A. W. and Olkin, I. (1997). A new method for adding a parameter to a family of distributions with application to the Exponential and Weibull families, *Biometrika*, 84(3):641–652. DOI: 10.1093/biomet/84.3.641. 9

[100] McCullagh, P. (1992). Conditional inference and Cauchy models, *Biometrika*, 79:247–259. DOI: 10.1093/biomet/79.2.247. 118

[101] Min, I. A., Mezic, I., and Leonard, A. (1996). Levy stable distributions for velocity and velocity difference in systems of vortex elements, *Physics of Fluids*, 8(5):1169–1180. DOI: 10.1063/1.868908. 129

[102] Nadarajah, S. and Kotz, S. (2007a). Multitude of beta distributions with applications, *Statistics: A Journal of Theoretical and Applied Statistics*, 41(2):153–179, DOI: 10.1080/02331880701223522. 53

[103] Nadarajah, S. and Kotz, S. (2007b). Skewed truncated Cauchy distribution with applications in economics, *Applied Economics Letters*, 14(13):957–961. DOI: 10.1080/13504850600705950. 128

[104] Norton, R. M. (1983). A characterization of the Cauchy distribution, *Sankhya-A*, 45:247–252. 121

[105] O'Hagan, A. and Leonard, T. (1976). Bayes estimation subject to uncertainty about parameter constraints, *Biometrika*, 63(1):201–203. DOI: 10.1093/biomet/63.1.201. 101

[106] Patil, G. P., Kapadia, C. H., and Owen, D. B. (1989). *Handbook of Statistical Distributions*, Marcel Dekker, NY. 103

[107] Patnaik, P. B. (1949). The noncentral chi-square and F-distributions and their applications, *Biometrika*, 46:202–232. DOI: 10.1093/biomet/36.1-2.202. 51, 109

[108] Pham-Gia, T. and Hung, T. L. (2001). The mean and median absolute deviations, *Mathematical and Computer Modeling*, 34(7):921–36. DOI: 10.1016/s0895-7177(01)00109-1. 9

[109] Pitman, J. and Yor, M. (1992). Arcsine laws and interval partitions derived from a stable subordinator, *Proc. of the London Mathematical Society*, 3-65(2):326–356. DOI: 10.1112/plms/s3-65.2.326. 66

[110] Pittman, E. J. G. and Williams, E. J. (1967). Cauchy-distributed functions of Cauchy variates, *Annals of Mathematical Statistics*, 38:916–918. DOI: 10.1214/aoms/1177698885. 123

[111] Quine, M. P. (1994). A result of Shepp, *Applied Mathematics Letters*, 7(6):33–34. DOI: 10.1016/0893-9659(94)90089-2. 94

[112] Rao, B. R. and Garg, M. L. (1969). A note on the generalized Cauchy distribution, *Canadian Mathematics Bulletin*, 12:865–868. 121

[113] Rao, C. R. (1965). On discrete distributions arising out of methods of ascertainment, *Classical and Contagious Distributions*, G. P. Patil (Ed.), Statistical Publishing Society, Kolkata, pages 320–333. 6

[114] Rao, C. R. (1973). *Linear Statistical Inference and its Applications*, 2nd ed., John Wiley, NY. DOI: 10.1002/9780470316436. 19, 50

[115] Rao, C. R. (1984). Weighted distributions arising out of methods of ascertainment, *Technical Report*, University of Pittsburg, PA, pages 38–84. https://apps.dtic.mil/dtic/tr/fulltext/u2/a145615.pdf DOI: 10.1007/978-1-4613-8560-8_24. 6

[116] Rather, N. A. and Rather, T. A. (2017). New generalizations of exponential distribution with applications, *Journal of Probability and Statistics*. https://www.hindawi.com/journals/jps/2017/2106748/ DOI: 10.1155/2017/2106748. 9

[117] Rider, P. R. (1957). Generalized Cauchy distributions, *Annals of the Institute of Statistical Mathematics*, 9:215–223. DOI: 10.1007/bf02892507. 121

[118] Roberts, C. (1966). A correlation model useful in the study of twins, *Journal of the American Statistical Association*, 61:1184–1190. DOI: 10.1080/01621459.1966.10482202. 101

[119] Roy, D. (2003). The discrete normal distribution, *Communications in Statistics—Theory and Methods*, 32(10):1871–1883. DOI: 10.1081/sta-120023256. 95

[120] Ryu, K. S. (2017). The arcsine law in the generalized analogue of Weiner space, *Journal of the Chungcheong Mathematical Society*, 30(1):67–76. 58

[121] Seber, G. A. F. (1963). The noncentral chi-squared and beta distributions, *Biometrika*, 50:542–44. DOI: 10.1093/biomet/50.3-4.542. 51

[122] Seber, G. A. F. (1965). Linear hypotheses and induced tests, *Biometrika*, 51:41–47. DOI: 10.1093/biomet/51.1-2.41. 50

[123] Shanmugam, R. (1991a). Significance testing of size bias in income data, *Journal of Quantitative Economics*, 7(2):287–294. 36

[124] Shanmugam, R. (1991b). Testing guaranteed exponentiality, *Naval Research Logistics Quarterly*, 38:877–892. DOI: 10.1002/nav.3800380607. 38

[125] Shanmugam, R. (1998). An index method of determining the sample size for normal population, *Communications in Statistics*, 27:1001–1008.

[126] Shanmugam, R. (2013). Tweaking exponential distribution to estimate the chance for more survival time if a cancerous kidney is removed, *International Journal of Research in Nursing*, 4(1):29–33. DOI: 10.3844/ijrnsp.2013.29.33. 38

[127] Shanmugam, R. (2014). Over/under dispersion sometimes necessitates modifying, *International Journal of Ecological Economics and Statistics*, 34(3):37–42. 114

[128] Shanmugam, R. (2017). $C(\alpha)$ expressions for split Gaussian frequency trend with illustration to check global warming, *International Journal of Ecological Economics and Statistics*, 38(3):23–42. 114

[129] Shanmugam, R. (2020). What do angles of cornea curvature reveal? A new (sinusoidal) probability density function with statistical properties assists, *International Journal of Data Science*, 5(1):53–78. DOI: 10.1504/ijds.2020.10031622. 82

[130] Shanmugam, R., Bartolucci, A., and Singh, K. (1999). Exponential model extended for neurological studies, *Epidemiology Health and Medical Research 2*, Oxley, L., et al. (Eds.), Modeling and simulation society of Australia Inc., pages 507–511. 38

[131] Shanmugam, R., Bartolucci, A., and Singh, K. (2002). The analysis of neurological studies using extended exponential model, *Mathematics and Computers in Simulation*, pages 81–85. DOI: 10.1016/s0378-4754(01)00395-0. 38

[132] Shanmugam, R. and Singh, J. (2012). Urgency biased beta distribution with application in drinking water data analysis, *International Journal of Statistics and Economics*, 9:56–82. 53

[133] Sharafi, M. and Behboodian, J. (2008). The Balakrishnan skew-normal density, *Statistical Papers*, 49:769–778. DOI: 10.1007/s00362-006-0038-z. 107

[134] Shepp, L. (1964). Normal functions of normal random variables, *SIAM Review*, 6:459. DOI: 10.1137/1006100. 94

[135] Singer, D. A. (2013). The lognormal distribution of metal resources in mineral deposits, *Ore Geology Reviews*, 55:80–86. DOI: 10.1016/j.oregeorev.2013.04.009. 114

[136] Spitzer, F. (1958). Some theorems concerning 2-dimensional Brownian motion, *Transactions of the American Mathematical Society*, 87:187–197. DOI: 10.1090/s0002-9947-1958-0104296-5. 63, 118

[137] Song, W. T. (2005). Relationships among some univariate distributions, *IEEE Transactions*, 37:651–656. DOI: 10.1080/07408170590948512. 71

[138] Steele, C. (2008). Use of the lognormal distribution for the coefficients of friction and wear, *Reliability Engineering and System Safety*, 93(10):1574–1576. DOI: 10.1016/j.ress.2007.09.005. 114

[139] Suksaengrakcharoen, S. and Bodhisuwan, W. (2014). A new family of generalized gamma distribution and its application, *Journal of Mathematical Statistics*, 10:211–220. DOI: 10.3844/jmssp.2014.211.220. 9

[140] Tio, G. G. and Guttman, I. (1965). The inverted Dirichlet distribution with applications, *Journal of American Statistical Association*, pages 793–805. DOI: 10.1080/01621459.1965.10480828. 51

[141] Whitson, C. H., Anderson, T. F., and Søreide, I. (1990). Application of the gamma distribution model to molecular weight and boiling point data for petroleum fractions, *Chemical Engineering Communications*, 96(1):259–278. DOI: 10.1080/00986449008911495. 78

[142] Williams, E. J. (1969). Cauchy-distributed functions and a characterization of the Cauchy distribution, *Annals of Mathematical Statistics*, 40(3):1083–1085. DOI: 10.1214/aoms/1177697613. 118

[143] Williams, M. A., Baek, G., et al. (2017). Global evidence on the distribution of GDP growth rates, *Physica A: Statistical Mechanics and its Applications*, 468(15):750–758. DOI: 10.1016/j.physa.2016.11.124. 129

[144] Winterton, S. S., Smy, T. J., and Tarr, N. G. (1992). On the source of scatter in contact resistance data, *Journal of Electronic Materials*, 21(9):917–921. DOI: 10.1007/bf02665549. 129

[145] Wise, M. E. (1950). The incomplete beta function as a contour integral and a quickly converging series for its inverse, *Biometrika*, 37:208–218. DOI: 10.1093/biomet/37.3-4.208. 51

[146] Yadegari, I., Gerami, A., and Khaledi, M. J. (2008). A generalization of the Balakrishnan skew-normal distribution, *Statistics and Probability Letters*, 78:1156–1167. DOI: 10.1016/j.spl.2007.12.001. 107

[147] Yalcin, F. and Simsek, Y. (2020). A new class of symmetric beta type distributions constructed by means of symmetric Bernstein type basis functions, *Symmetry*, 12(5):779–790. DOI: 10.3390/sym12050779. 9, 42

[148] Zaninetti, L. (2010). The luminosity function of galaxies as modelled by the generalized gamma distribution, *Acta Physica Polonica B*, 41:729–751. 76

[149] Zaninetti, L. (2019). New probability distributions in astrophysics: I. The truncated generalized gamma distribution, *International Journal of Astronomy and Astrophysics*, 9(4), preprint. https://arxiv.org/abs/1912.12053 DOI: 10.4236/ijaa.2019.94027. 76

Authors' Biographies

RAJAN CHATTAMVELLI

Rajan Chattamvelli is a professor in the School of Advanced Sciences at VIT University, Vellore, Tamil Nadu. He has published more than 22 research articles in international journals of repute and at various conferences. His research interests are in computational statistics, design of algorithms, parallel computing, data mining, machine learning, blockchain, combinatorics, and big data analytics. His prior assignments include Denver Public Health, Colorado; Metromail Corporation, Lincoln, Nebraska; Frederick University, Cyprus; Indian Institute of Management; Periyar Maniammai University, Thanjavur; and Presidency University, Bangalore.

RAMALINGAM SHANMUGAM

Ramalingam Shanmugam is a honorary professor in the school of Health Administration at Texas State University. He is the editor of the journals *Advances in Life Sciences*, *Global Journal of Research and Review*, and *International Journal of Research in Medical Sciences*, and book-review editor of the *Journal of Statistical Computation and Simulation*. He has published more than 200 research articles and 120 conference papers. His areas of research include theoretical and computational statistics, number theory, operations research, biostatistics, decision making, and epidemiology. His prior assignments include University of South Alabama, University of Colorado at Denver, Argonne National Labs, Indian Statistical Institute, and Mississippi State University. He is a fellow of the International Statistical Institute.

Index

Printed in the United States
by Baker & Taylor Publisher Services